Developing Integrated Communications Strategy

Postgraduate Certificate in Global Marketing Practice

Developing Integrated Communications Strategy

First edition January 2010

ISBN: 9780 7517 8410 7

British Library Cataloguing-in-Publication Data
A catalogue record for this book has been
applied for from the British Library

Published by

BPP Learning Media Ltd
Aldine House, Aldine Place
London W12 8AA

www.bpp.com/learningmedia

Printed in Great Britain

We are grateful to the GMN Advisory Council
members Crispin Reed, Damon Segal, Leslie
de Chernatony and David Haigh.

Author:
Dr Kellie Vincent with guest chapter from
Damon Segal

A note about copyright

Dear Customer

What does the little © mean and why does it
matter?

Your market-leading BPP books, course materials
and e-learning materials do not write and update
themselves. People write them: on their own
behalf or as employees of an organisation that
invests in this activity. Copyright law protects their
livelihoods. It does so by creating rights over the
use of the content.

Breach of copyright is a form of theft – as well as
being a criminal offence in some jurisdictions, it is
potentially a serious breach of professional ethics.

With current technology, things might seem a bit
hazy but, basically, without the express permission
of BPP Learning Media:

- Photocopying our materials is a breach of
  copyright

- Scanning, ripcasting or conversion of our
  digital materials into different file formats,
  uploading them to facebook or emailing
  them to your friends is a breach of
  copyright

You can, of course, sell your books, in the form in
which you have bought them – once you have
finished with them. (Is this fair to your fellow
students? We update for a reason.) But the e-
products are sold on a single user licence basis:
we do not supply 'unlock' codes to people who
have bought them second-hand.

And what about outside the UK? BPP Learning
Media strives to make our materials available at
prices students can afford by local printing
arrangements, pricing policies and partnerships
which are clearly listed on our website. A tiny
minority ignore this and indulge in criminal activity
by illegally photocopying our material or supporting
organisations that do. If they act illegally and
unethically in one area, can you really trust them?

Contents

Congratulations

on choosing to study towards MSc Global Marketing Practice.

You have made the right decision because you have at your fingertips a wealth of support to help guide you through your studies with BPP Learning Media, Ashcroft International Business School and Global Marketing Network.

GMN's CEO Darrell Kofkin summarises the key value of The Global Marketer Programme as

> *"The programme is designed to prepare you thoroughly for a rewarding career in marketing and ensure you can make a profitable contribution to business anywhere in the world."*

For the first time, you can achieve an academic award at Masters level from one of the UK's largest and most progressive universities, while at the same time ensuring you meet all the academic criteria required for acceptance as a Professional Member of Global Marketing Network.

It is our intention that at this first level of this innovative Masters programme, Postgraduate Certificate Global Marketing Practice, you are able to develop the skills you require to perform effectively from day one.

This Study Text is designed to work in conjunction with the following elements of the overall programme:

- A Student Handbook designed to give you all the information you need to achieve success

- A personal E-Tutor to guide, encourage and develop your thinking and provide guidance for your assignment

- Access to The Global Marketer virtual learning environment providing you with a week-by-week plan of study for each module, with access to articles, case studies, additional learning activities and a global discussion forum connecting you with other programme participants

- Access to the Anglia Ruskin University digital library, providing access to an extensive range of online databases, world leading journals and E-books

- Access to the GMN online network with all the latest news about approved conferences, workshops and events that you can attend to support your continuing development

Each chapter of the Study Text is designed to refer to each Study Session within the programme.

Integrating Global Marketing Network Philosophies

Each module has at least one GMN Advisory Council or Faculty member acting as Module Advisor. Module Advisors contribute to the development of the curriculum and study materials.

The GMN Module Advisor for Developing Integrated Communications Strategy is Crispin Reed. **Crispin** is Managing Director of Brandhouse, a global brand design and strategy consultancy with a full range of multinational clients and often commentates in the media on marketing issues.

Discussing the programme, Crispin Reed stated:

> *"This new programme has been genuinely designed for the global marketer of the 21st century. The development process has been robust, rigorous and forensic. I have no doubt it will equip practitioners with the necessary skills to rise to the very top of the profession."*

Additional information and reviews of the Study Materials by Module Advisors can be found on the Global Marketer Programme supporting Virtual Learning Environment.

Using this Study Text

This Study Text includes features designed specifically to make learning effective and efficient.

- Each chapter begins with an **introduction** to set the chapter in context.

- Chapter **learning outcomes** summarise what you are expected to be able to achieve through reading the chapter. The content covered is summarised in a Chapter Roundup which is organised according to these learning outcomes at the end of the chapter.

- Harvard referencing is used throughout the chapter with **a full reference list** at the end of each chapter.

Throughout the Study Text, there are also special aids to learning. These are indicated by the following symbols.

Definition

Definitions are provided for key terms.

Activity 1

Activities for you to complete are included througout the chapters. Often these will ask you to relate theories to your own organisation. Where appropriate activitiy debriefs are included at the end of the chapter.

Global case study

Real life global examples are used to illustrate marketing practice.

 Readily available key text book resources, significant chapters and additional reading are selectively added to the chapters to help develop your further reading.

 Selected online materials which link to the topics covered within the chapter are suggested at appropriate points.

Assessment advice

Points to bear in mind in relation to specific chapter topics when preparing your assignment are inserted at relevant places in the chapter.

GMN viewpoint

GMN experts' opinions, snippets from presentations and papers are included to bring both practitioner and academic cutting edge perspectives into the material.

About this module

This module is work-based in its nature and learning approach and builds on your existing practice-based awareness of branding, digital marketing and communication theories, practices and issues arising. It supports the development of relevant employability, promotability and professional skills that enable you to develop an integrated communications strategy for an organisation that enhances the value of the brand.

The module will advance your understanding of these issues by providing a critical and, where necessary, targeted insight into the latest strategic brand management, digital marketing and communication theories and practices so that you can develop, recommend, implement and monitor an integrated communications strategy for an organisation deploying both traditional and non-traditional media. Where possible, the emphasis will be placed on your own organisation. Three key areas are covered:

1. Strategic brand management

The module will enable students to develop effective branding strategies drawing upon the theories of branding coupled with practical examples of what works and why, and what results have been achieved, especially in terms of calculation and improvement in brand value and brand alignment.

2. Developing an effective digital marketing strategy

The module will enable students to explore the use and application of current digital marketing tools (eg email, mobile marketing [ie SMS, MMS], websites, Search Engine Optimisation, social networking communities, pay-per-click and other online revenue models, viral marketing etc.).

We draw upon the latest models and theories of digital marketing and will include further practical examples of what works and why, and what results have been and could be expected to be achieved.

3. Developing and implementing an effective integrated communication strategy

Drawing upon the well-established and proven theories and models of communication, the development and use of communication tools and processes will be investigated with reference to practical case studies and examples. Particular attention will be paid to how clear, concise and cogent messages can be developed for the target audience(s) together with the levels of awareness, understanding and acceptance that could be expected as a consequence of the deployment of the communication strategy. You will be able to critically evaluate and recommend effective monitoring and control systems which measure the performance of the organisation's integrated communications strategy and the importance of considering the legal issues in its implementation.

Module aims

You are expected to demonstrate an understanding of the brand management, digital marketing and integrated communications strategy issues facing the marketer, rooted in proven models and concepts as set out in the marketing strategy literature provided/recommended.

You will be assessed via an assignment related either to your own organisation or a national/international business organisation. This will ensure a clear integration between theory and practice. In order to complete this assignment, you should draw upon a range of theories and models ranging from classical theories to contemporary views of strategy including emergent, collaborative and ethical perspectives across the Brand management, Digital Marketing and Communications disciplines.

The module supports the development of relevant employability, promotability and professional skills which enables the student to develop an understanding of the methods and approaches necessary to develop, implement and monitor an effective integrated communications strategy organisation for the organisation.

Focussing on your own organisation (or one that you are familiar with or a case study) facilitates the application of, and reflection on, theory in practice.

A note on pronouns

On occasions in this Study Text, 'he' is used for 'he or she', 'him' for 'him or her' and so forth. While we try to avoid this practice it is sometimes necessary for reasons of style. No prejudice or stereotyping accounting to sex is intended or assumed.

Key contacts

Please email gmn@bpp.com for any queries about this Study Text and The Global Marketer Programme.

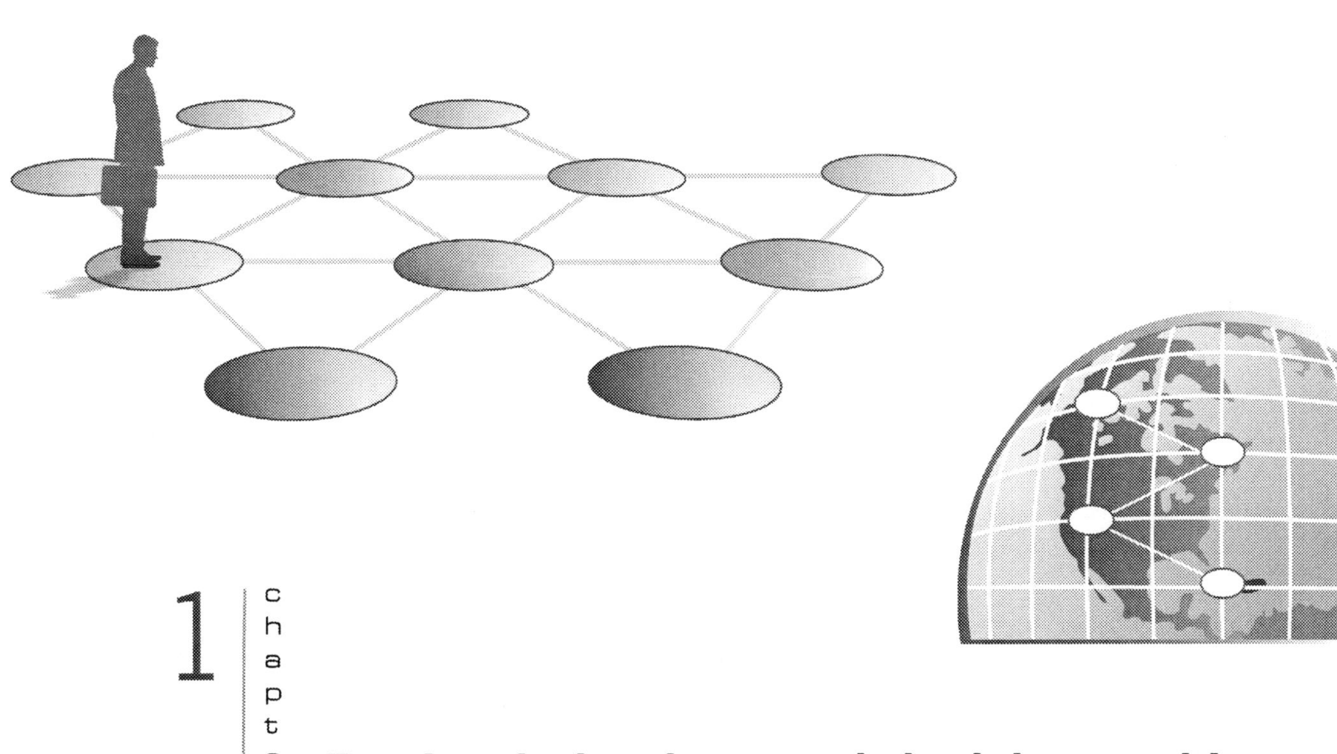

1 chapter

Buying behaviour and decision making

We begin our first chapter by considering the nature of buyers and their purchase behaviour.

When we discuss marketing communications, we are really referring to how an organisation *communicates* with a large number of stakeholders. As with any communication, the success of transferring the *intended* message will be dependent on how both parties interpret actions and behaviours. Interpretations are made according to previous personal experiences and the context of individuals' lives. In any marketing communications strategy, the first step should therefore be to identify, analyse and understand the target market and buying behaviours exhibited.

The aim of this chapter is to provide an overview of buyer behaviour in readiness to apply the concepts when we move on (in later chapters) to consider the specific aspects of an integrated marketing communications strategy. Brands represent different things to different groups of consumers and so understanding how their perceptions and attitudes influence their behaviour is vital in order to integrate successful brand (and digital) strategies.

Contents

Together with the session activities to complete on the Virtual Learning Environment, this chapter will enable you to:

1   Introduce the concept of buyer behaviour and state why this is important to communications
2   Consider how communication works
3   Examine the purchase decision making process
4   Evaluate models of buyer behaviour
5   Appreciate the importance of decision making units
6   Consider the principles of buyer motives

1   The relevance of buyer behaviour to integrated marketing communications

The study of consumer buying behaviour by an organisation is important for a number of reasons.

(a)   The **buyer's reaction** to the organisation's marketing strategy has a major impact on the success of the organisation.

(b)   If organisations are truly to implement the marketing concept, they must examine the main influences on **what, where, when and how customers buy**. Only in this way will they be able to devise a marketing mix that satisfies the needs of the customers.

(c)   By gaining a better understanding of the factors influencing their customers and how their customers will respond, organisations will be better able to predict the **effectiveness of their marketing activities**.

However, not all consumers behave in the same way. Decision making and purchase patterns of behaviour vary considerably between individuals and across product categories. This **complexity must be recognised** and **marketing communications adapted** to meet the needs of different customers.

As we work through this module throughout the remaining Study Text chapters you will be building upon the theories of communication covered within this chapter.

Marketing communications involves all communications by an organisation and the various stakeholders who might be involved are not just its customers.

Marketing communications is one of the elements of the marketing mix, and is responsible for the promotion of the marketing offer to the target market. We must continually remind ourselves that we are not studying marketing communications strategies in isolation from the rest of the business or organisation. If they are to be successful, **marketing communications strategies must be co-ordinated with the overall planning process**. This is a very broad perspective and before we go any further we need to briefly consider the tools and methods of marketing communications.

The following table adapted from De Pelsmacker, Geuens and Van den Berg (2007) highlights the variety of marketing mix instruments. The tools available within the communications mix (which are important to this module) are shown within the '*Promotion*' column.

Product	Price	Place	Promotion (Communications mix)
Benefits	List price	Channels	Advertising
Features	Discounts	Logistics	Public relations
Options	Credit terms	Inventory	Sponsorship
Quality	Payment periods	Transport	Sales promotions
Design	Incentives	Locations	Direct marketing
Branding			Point of purchase
Packaging			Exhibitions and trade fairs
Services			
Warranties			Personal selling Internet

The promotional tools represent the deployment of **deliberate and intentional methods** calculated to bring about a favourable response in the customer's behaviour. The *'Promotion'* column within the table represents the most obvious communication methods, though other parts of the marketing mix, including the product itself, pricing, policy and distribution channels, will also have decisive effects.

Choosing the correct tools for a particular promotions task is not an easy one. The process is becoming **increasingly scientific** because of access to consumer and media databases as well as the high level of information available as a result of online interactions. Matching consumer characteristics with media characteristics can be rapidly carried out however, the expertise of the marketing manager is vital.

Marketing and marketing communications are changing at a rapid pace. No longer are marketers managing a specific marketing function (for example advertising and promotion). The marketing role boundaries are now blurred and all marketers need to understand the complex networks of relationships within and outside their organisation to identify how to best communicate and maintain positive relations with many stakeholders. The change in focus from customer acquisition to building long lasting relationships with selected customers while maintaining wellbeing for society within a highly interactive global marketplace is now a reality.

In recent years there has been a recognition that in order for communication to be effective, it is essential to co-ordinate a consistent and integrated approach. Integrated marketing communications recognises the need to design all relevant communication messages together so that they work in harmony, consistently and efficiently (Smith & Taylor, 2004).

Marketing communications, regardless of the tools employed, can be viewed partly as an attempt by an organisation or brand to create and sustain dialogue with various stakeholders. Communication itself is the process by which individuals share meaning and so in order for dialogue to occur, each participant in the communication process needs to be able to interpret the meaning embedded in the messages (Fill, 2009). Ultimately, the tools and techniques used to communicate (either **directly** as is the case with typical communications such as advertising and new media or **indirectly** as with branding) should be designed and used based on an understanding of how their properties will impact on the **behaviour** of the target audience (who are often buyers). An understanding of behaviour is therefore a prerequisite to understanding how marketing communication works.

2 Understanding how communication works

Marketing communications can be considered in the context of the **exchange process**. Exchanges, or transactions, are central to much modern thinking in economics. All activity undertaken by individuals or organisations can be seen as a series of exchanges.

Communication is central to the exchange process. The DRIP model is often used to explain this point.

D   **Differentiates** between competing offerings, helping consumers to decide which exchanges to make, helping to prevent monopolies from developing and encouraging lower prices.

R   **Reminds** customers of the benefits of past transactions and so convinces them that they should enter into a similar exchange.

I   **Informs** and makes potential customers aware of an organisation's offering.

P   **Persuades** current and potential customers of the desirability of entering into an exchange relationship.

We will briefly discuss relevant theoretical models of communications because this will help you later when you think about the way in which communications are interpreted by customers and acted upon during a purchase scenario.

2.1 Communication theory models

If you ask somebody the time and get a response, or send an invoice and receive payment, this corresponds to a single cycle of communication.

A model of the communication cycle (known as the **'radio signal' model** (Schramm, 1955)) can be depicted as follows.

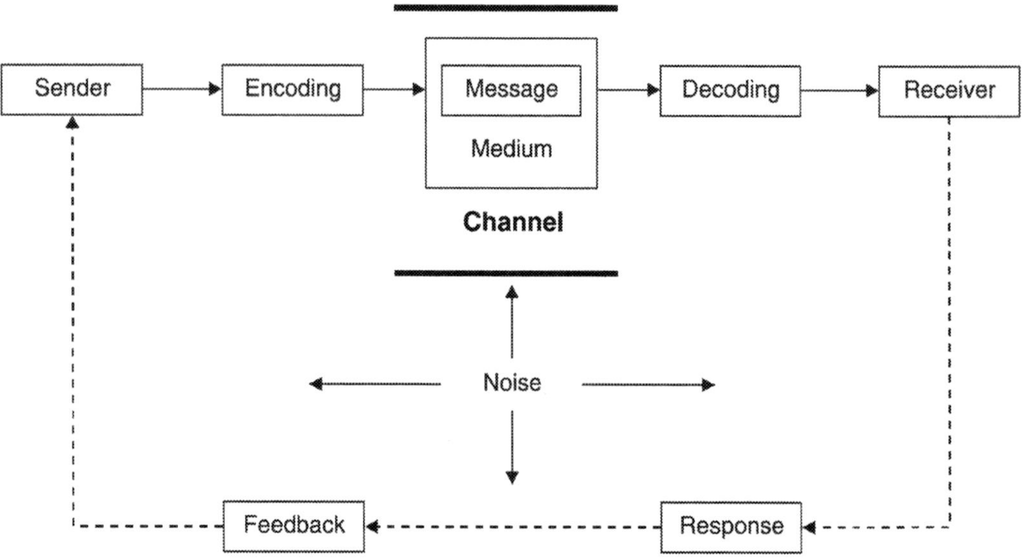

The 'radio signal' model of communication

2.1.1 Planning the message

The **sender** decides to communicate and plans what and how to communicate. In written communications (eg drafting an advertisement or report) you have 'off-line' decision and planning time in which to work out how best to present your ideas to get the desired response from your target audience. In face-to-face and oral communication (eg answering a telephone or customer query) you may have to plan and formulate messages 'live'. This is why oral and face-to-face communication skills are so important.

2.1.2 The encoding analogy

Encoding is an analogy for how the sender puts his or her intentions into a form which can be transmitted: a form which the sender and receiver must *both* understand, if the sender's message is to be correctly interpreted at the other end by the receiver. This is like a code, because the words, numbers, pictures and gestures we use are only symbols representing our ideas. In communicating we must translate our ideas into a code which we think the receiver will be able to 'decipher' or translate back into the ideas.

It is important to bear in mind that a symbol that we use and understand may be ambiguous (have more than one possible meaning) or mean something different to a person of different age, nationality, experience or beliefs. Just because we understand what we mean, it does not necessarily follow that someone else will: another reason to consider the needs and capacities of the target audience. Never mind what you think you'll say; what do you think *they'll hear*?

2.1.3 Media and the channels

Once the idea has been encoded as a message, the sender needs to choose how to **transmit** or get it across to the receiver/target audience.

(a) The particular route or path via which the message is sent, connecting the sender and receiver, is called the **channel** of communication: examples include a notice board, postal/telecommunications system and the Internet.

(b) The tool or instrument which is used is the **medium** (plural 'media'). The selected medium will usually come under one of the broad headings of:

- **Visual** (non-verbal) communication: a gesture, chart, graphic poster or Web page

- **Written** (verbal) communication: a letter, memo, report, press-release, email or text advertisement

- **Oral** (verbal) communication ('by mouth'), which includes both face-to-face media (meetings, presentations) and remote media (telephone, television)

The choice of medium is an important communication skill: we will discuss it further later in the chapter.

If everything has gone well so far, the **receiver** receives the message – and must then **decode** its meaning, using his or her own knowledge, skills and perceptions to interpret its meaning.

2.1.4 Feedback

Definition

Feedback is a response of the receiver which indicates to the sender that the message has (or has not) been successfully received, understood and interpreted.

Seeking and giving **feedback** (especially in oral and face-to-face communication) is a key communication skill. Communication is prone to a wide range of factors which interfere with the effective transmission of

the message from sender to receiver. These factors are often known by the collective term **'noise'**. They include technical interference (such as a bad phone line or poorly printed advertisement) and sources of misunderstanding (such as emotion, prejudice and differences in perception) and so on. Only by seeking feedback can the sender trust that the message has been both received and understood.

Examples of **positive feedback** include: a staff member accurately reading back your instructions; a presentation audience nodding agreement; a customer sending back the reply form as requested; your target audience buying the advertised product – or, as an example of more systematically gathered feedback, a market research focus group confirming that their response to a brand is as you expected. (You can probably work out what the negative equivalents would be.)

Think of a recent piece of communications within your organisation and break down its components using the radio signal model of communication.

2.1.5 Noise

This communication process is **not carried out in isolation**. There are many senders competing for the attention of the receiver. As a result there is considerable noise in the environment and an individual may be bombarded by several hundred commercial messages each day.

2.1.6 Realms of understanding

Fill (2009) reports that successful communications are more likely to be achieved if the source and the receiver understand one another. Understanding is related to attitudes, perceptions, behaviour and experience which together form the values of both parties. Common ground is therefore required for communication to be successful. Understanding the target market (audience) and testing communication messages is employed in order to understand the receivers of messages and more effectively construct messages that will be decoded to positive effect.

2.1.7 Influences of the communication process

Fill (2009) argues that there is unlikely to be a single model of communication which accurately reflects all forms of communications. He states that in any model there should be a consideration of two influences on the communication process:

Influence 1   **Media used to convey information**

Influence 2   **People**

2.1.8 Influence of media

The developments in new media in recent years has led to the development of new theories of media because it is becoming more widely available, faster and less easy to control than traditional media. New media and digital technologies provide an opportunity for increasing levels of interaction and dialogue between marketers and their audiences. Likewise, they also enable interaction and dialogue between different stakeholder audiences, dialogue that the marketer has limited input and no control over.

Media Richness Theory was however developed before the influence of the Internet (Fill, 2009) and considers the depth of message content and the capability of different media to process ambiguous communication and promote understanding. The criteria used to assess the richness of each media type is covered below.

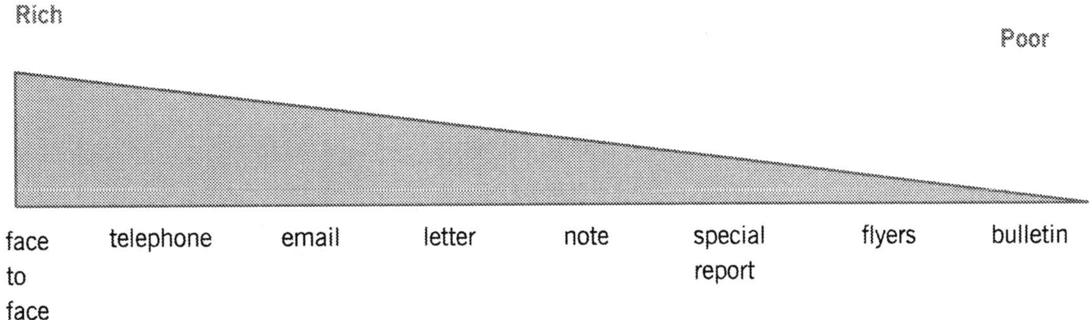

Criteria used to assess 'richness'
Ability to provide instant feedback
Capacity to transmit multiple communication cues such as tone of voice, body language, eye contact
Use of natural language
Personal focus

Face to face communication is the richest of all communication media and helps to establish a personal connection. Numeric and formal written communication conversely is slow and so feedback is delayed and visually limited (Fill, 2009).

The spectrum of richness can be depicted as:

Digital media is expected to be relatively rich because it enables a high degree of interactivity and enables the simultaneous delivery of messages in a variety of formats.

2.1.9 Influence of people

There is not just one receiver, but many, and the receivers may communicate with each other regarding the sender's message. By virtue of experience or social standing one particular receiver may influence the opinions of others regarding the product that is the subject of the message. This is known as **opinion leadership** and may lead to a **two-step flow of communication**, whereby mass messages are filtered through opinion leaders to a mass audience.

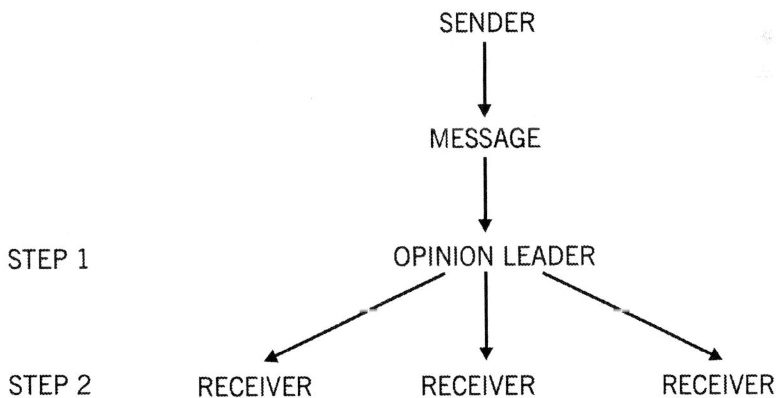

Opinion leaders are 'people within a reference group who, because of special skills, knowledge, personality traits or other characteristics, exert influence on others' in regard to a particular product or decision area (Kotler et al, 1999, p.98).

Later in the chapter we will look at a model of consumer behaviour known as the **Diffusion of Innovation**. The model refers to the process and how quickly consumers start to purchase and use new products and services. The impact of the first 'innovator' consumers, and indeed their reviews of the new products, is to be important 'informants' or opinion leaders for future consumers. Word of mouth has for a long time been recognised as a highly influential way to change individuals' behaviour. Taxi drivers in a number of countries, including the UK and India, have been employed by some tourism experts to discuss the benefits of particular locations as holiday destinations with their passengers. The benefit to the drivers is

that in order to be able to speak with conviction they are often taken to enjoy the destinations for themselves first.

In today's world where global reviews are readily available, because of the proliferation of online review websites, the influence of people is even more relevant.

Tata Nano

The Tata Nano has been described by some media sources as a completely brand new category of car. Launched in March 2009, it was initially designed for the Indian market. The first car was sold on 17th July 2009 from a dealership in Mumbai. Speaking on the occasion, Mr. Tata said,

"I hope the Tata Nano will bring motoring pleasure to those who will be buying their first car as also those who currently own cars but want a modern, contemporary, emission-friendly city car." (Tata, 2009).

The car is smaller, eco-friendly (electric models to be made available), considerably cheaper and more customisable than other cars in any category. As a result the launch led to much interest throughout the world.

Driver reviews and motoring expert opinion leaders (often the motoring press) are always important influencers to car buyers. In the case of the Nano, because it is such a different proposition, the opinions of the *'early adopter'* buyers and test drivers will be particularly influential.

Tata were prepared for this and set up their own Facebook page in order to contribute to discussions which would have taken place between potential buyers and other interested stakeholders anyway. They also report any positive press reports on their website and use the web to answer detailed questions about this new design of car outlining the related benefits to consumers.

Take a look at the Tata Nano website:

http://tatanano.inservices.tatamotors.com/tatamotors/

and specifically use the links to look at the press coverage and Facebook sites.

2.2 Perception

Definition

Perception can be defined as a process by which people select and interpret stimuli into a meaningful picture.

The way consumers view an object could include their mental picture of a brand, or the traits they attribute to a brand. The way that a person perceives a situation will affect how they act. Possible differences in perception can be explained by three perceptual processes.

- Selective attention
- Selective distortion
- Selection retention

2.2.1 Selective attention

A receiver will not notice all the commercial messages that he comes into contact with, so the sender must design the message so as to **win attention** in spite of the surrounding noise. Repetition, size, contour, music and sexual attraction are features used to attract attention.

2.2.2 Selective distortion

In many cases receivers distort or change the information they receive if that information does not fit in with their existing beliefs. In other words, people hear what they want to hear. Selective distortion may take a variety of forms.

- Amplification (where receivers may add things to the message that are not there)
- Levelling (where receivers do not notice other things that are there).

The task of the sender is to produce a message that is clear, simple and interesting, what many refer to as **likeable**.

2.2.3 Selective recall and message rehearsal

A receiver will retain in memory **a small fraction of the messages** that are perceived and processed. The sender's aim, therefore, is to get the message into the receiver's **long-term memory**, because once in the long-term memory the message can modify the receiver's beliefs and attitudes. However, to reach the long-term memory the message has to enter the short-term memory, which has only a limited capacity to process information. The factor influencing the passage of the message from the short-term to the long-term memory is the amount and type of **message rehearsal** given by the receiver.

In message rehearsal the receiver elaborates on the meaning of the message in a way that brings related thoughts from the long-term memory into his short-term memory.

(a) If the receiver's initial attitude to the object of the message is positive and he rehearses support arguments then the message is likely to be accepted and have high recall.

(b) If the receiver's initial attitude is negative and the person rehearses counter arguments against the object of the message then the message is likely to be rejected, but it will remain in the long-term memory.

2.3 Learning

Learning concerns the process whereby an individual's behaviour changes as a result of their experience. Theories about learning state that learning is the result of the interplay of five factors.

- Drives
- Stimuli
- Cues
- Responses
- Reinforcement

(a) **A drive is a strong internal force impelling action**, which will become a motive when it is directed to a particular drive-reducing **stimulus** object (the product).

(b) **Cues are minor stimuli** (such as seeing the product in action, favourable reactions to the product by family and friends) that determine when, where and how the person responds.

(c) If the purchase experience is rewarding, then the **response** to the product will be **reinforced**, making a repeat purchase the next time the situation arises more likely.

2.4 Beliefs and attitudes

> **Definition**
>
> A **belief** is a thought that a person holds to be true about something.

Beliefs are important to marketers as the beliefs that people have about products make up the brand images of those products.

> **Definition**
>
> An **attitude** describes a person's 'enduring favourable or unfavourable cognitive evaluations, emotional feelings, and action tendencies toward some object or idea' (Kotler et al, 1999).

Attitudes lead people to behave in a fairly consistent way towards similar objects. Attitudes can be regarded as a short-cut in the thought process by ensuring that people do not have to interpret and react to every object in a fresh way. Attitudes settle into a consistent pattern, and to change one attitude may entail major changes to other attitudes. Attitude can also be considered as a learned tendency to respond to something in a consistently positive or negative manner .... Attitude is an evaluation: what we feel about a concept (brand, category, person, ideology and so on).

2.4.1 How to change attitudes

In changing attitudes, we can change:

- Cognition
- Affectivity
- Conation

> **Definitions**
>
> **Cognition**: what people think about a product.
>
> **Affectivity**: what people feel about a product.
>
> **Conation**: what people actually do; their behaviour.

2.4.2 Changing the way people think

With regard to cognition, the aims of the marketing communicator would be:

- To change beliefs entirely
- To change the relative importance of existing beliefs
- To develop new beliefs

Messages that change cognition should then have a 'knock on' effect on the other two components, leading to an overall change in attitude:

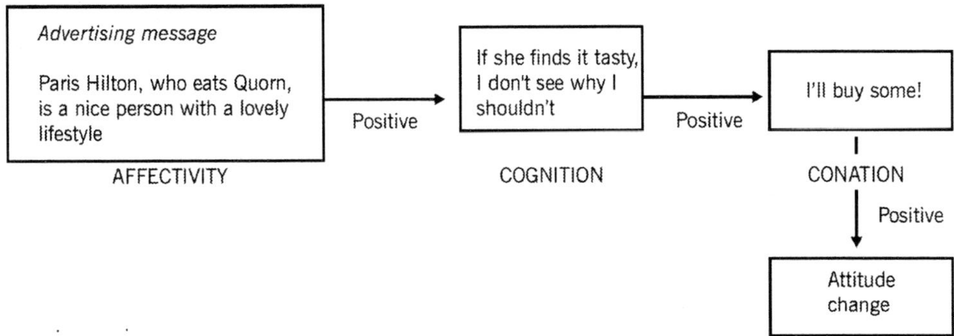

2.4.3 Changing the way people feel

Marketers can attempt to influence customers' feelings (**affectivity**) directly, before touching either beliefs or behaviour. The idea is to make the experience of using the product seem an enjoyable one (or a worthy, or exciting one).

There are three main ways in which a message can change affectivity.

(a) **Repetition** of positive messages when the product is one in which the consumer is not much involved (eg washing up liquid).

(b) **Likeable advertising** means that the consumer is more inclined to like the product.

(c) **Classical conditioning** works by linking the product name with a stimulus that is liked by the consumer, such as music.

2.4.4 Changing the way people behave

Often a consumer will try or **experiment** with a product before either his beliefs or his feelings are changed, most usually because a product with fairly **low involvement** is **reduced in price**.

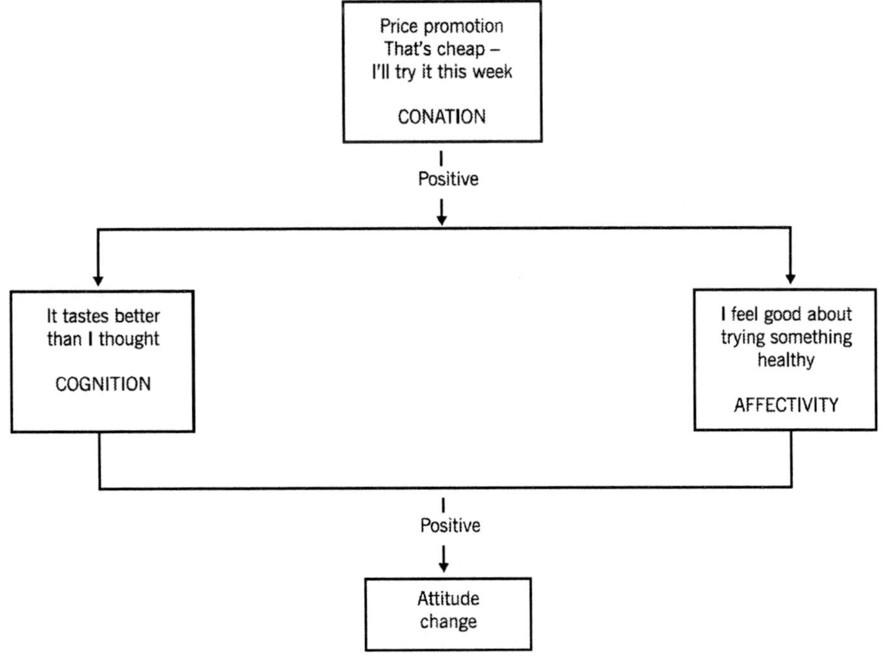

Generally, the behaviour marketers seek to change is for target consumers to identify their brands as ones in which they would recommend or purchase for themselves either now or in the future.

3 Purchase decision making

3.1 Customers and markets

When we think about buyers, we need to be very careful about terminology because whether somebody is a customer or a consumer, buyer or a payer will influence their purchase behaviour.

> **Definitions**
>
> - **Customers** are people who use goods and services and **pay** for them.
>
> - **Users** are people who use the product or service, but **do not pay** for it.
>
> - A **consumer** is the end user of a product or service who may or may not be the customer. For example, the consumer of a tin of cat food will be the family pet, while it was the pet owner who was the customer.
>
> - The **payer** is the person who finances the purchase.
>
> - The **buyer** participates in the procurement of the product from the market-place.

Not all organisations accept this distinction between **'customer'** (the person who **pays**) and **'user'** (the person who **consumes**). In particular, **not-for-profit organisations** such as local authorities, libraries and schools often have a diversity of people and groups interested in the products and services they supply. These people and groups may be variously described as 'clients', 'users' or 'customers'.

The roles can be combined in many different ways.

User's role		Example	
Payer	**Buyer**	**Business**	**Consumer/domestic**
No	No	• Furniture, eg equipment used by employees but procured	• Cat food • Car purchased by parents for sole use of child • Health insurance policy paid by the company
Yes	No	Office manager may buy and pay for stationery for use by others	Stockbrokers are agents who buy shares on behalf of the user
No	Yes	Business-related examples: corporate sponsorship can mean that universities can select a building contractor and enjoy the subsequent benefits of the builder's work (in the form of, say, a dedicated business school or residential conference centre) without having to meet the costs.	Consumer-based examples: the driver who selects a car breakdown service with the costs reimbursed by an insurance company; the bride who specifies prospective wedding gifts from a 'list' maintained through a retail operation.
Yes	Yes	Small-business entrepreneurs often combine all three roles when they acquire office equipment, furniture, and the services of an accountant.	Most consumers purchase and pay for products intended for their personal use, such as clothing, watches, airline tickets, haircuts, and so forth.

 Developing Integrated Communications Strategy

3.1.1 Reasons for role specialisation

Users are unlikely to play other customer roles when the:

User lacks:	**Product is:**
• Expertise/knowledge to make a choice	• Unaffordable
• Time to evaluate choice	• Subsidised (eg staff canteen)
• Buying power	• Free (eg public library)
• Access to the product	

Distinctions between user, payer and buyer are important because their clarification of the buying process helps potential product/service suppliers to **focus their marketing efforts** to optimal advantage. On the other hand, these same distinctions, even when clarified, do not necessarily show where (or how) power and influence are exercised, and by whom.

Global case study

In a British family, it is still mostly women who are the main food shoppers. Some food items however are clearly targeted at men or children. A pie manufacturer introduced a new range of reduced size mini pork pies. In the British market these are often eaten as an informal party snack and are regarded generally as a traditional but high fat product. Research for the pie manufacturer found that their target male typically ate whole pies in the larger size intended to be shared (and often just as a snack). The new smaller sized pork pies were cleverly positioned to appeal to female food shoppers as well as the male consumers who were likely to eat them.

The smaller size was specifically developed to meet the needs of women who were concerned about the high levels of fat their male partners were eating but didn't want to stop buying a favourite food.

Activity 2

Try to think of some more examples where the distinction between customer and user is significant.

3.2 Types of consumer buying decisions

On a continuum of effort ranging from very high to very low, three specific levels of consumer decision-making can be distinguished.

3.2.1 Limited problem-solving

At this level of problem-solving, consumers already have established the basic criteria for evaluating the product category and the various brands in the category. However, they have not fully established preferences concerning a select group of brands. Their search for additional information is more like 'fine tuning', they must gather additional brand information to discriminate among the various brands, or within brand families but between products with differing features, for example, when buying a new TV.

3.2.2 Routinised response behaviour

At this level, consumers have experience with the product category and a well-established set of criteria into which to evaluate the brands they are considering. In some situations, they may search for a small amount of additional information, in others they simply review what they already know, for example when buying breakfast cereal.

3.3 Types of organisational buying decisions

The level of problem-solving in an organisational buying decision will also vary in its complexity.

Some purchases will be a **straight rebuy** or the routine topping up of stocks without changing supplier or product specifications: for example, the re-ordering of stationery supplies. (This may be done on an automatic re-ordering system by the purchasing department or even the supplier, requiring only post-purchase review to ensure satisfaction.)

Some purchases will be a **modified rebuy**: the organisation wants to change product specifications, prices, terms or suppliers (which should stimulate competitive offerings from existing and alternative suppliers). Any or all stages of the buying decision may be revisited.

Some purchases will be a **new task** situation: the organisation is buying a product or service for the first time. In such circumstances, an extensive and systematic decision-making process may take place. This is an opportunity for the marketer to reach key members of the buying centre and to offer support and information in making the decision.

3.4 Involvement theory

Often we say that the purchase scenario is high involvement or low involvement. By involvement we mean how important and relevant the product or purchase context is to the consumer.

> **Definition**
>
> **Involvement theory**
>
> The level of care an individual experiences when considering the purchase of products or services (Fill, 2009).

The roots of involvement theory stem from social psychology and according to Kapferer and Laurent (1985) there are five different elements which determine whether a high or low level of involvement will exist:

The authors reported that:

"[To summarise] five antecedent conditions of involvement have been used by researchers: interest in the product category, enjoyment or pleasure derived from it, perception of self expression through product category, and the two components of perceived risk: the stake and subjective probability factors. The first facet (interest) is an antecedent of enduring involvement only. Pleasure and sign value may apply to both enduring and situational involvement. Perceived risk induces mostly situational involvement."

This is quote is deconstructed in the following list.

1. **Interest in the product category** – the level of interest and relevance in the product type
2. **Hedonic value** – pleasure in the purchase
3. **Sign value** – ability to project self expression through association with the product category
4. **Risk importance** – in terms of performance, financial, physical, social, time and ego related risk
5. **Perceived risk probability** – expected likelihood of a negative outcome occurring

The table below outlines risk in more detail and how marketers can attempt to deal with it.

Risk	Comment	Dealing with it
Performance risk	Will the product function properly?	Guarantees money back
Financial risk	Can I afford it, is it good value?	Emphasise value for money, quality
Physical risk	Will the product harm me or other people?	Emphasise safety
Ego risk	Will the product satisfy my needs for self esteem and self image?	Aspiration groups
Social risk	Will significant others disapprove?	Suggest psychological rewards
Time risk	Have I the time to go shopping to buy this product?	Importance or convenience of product

High involvement will occur when the consumer perceives that there is high level of risk and personal relevance associated with the purchase. Typical products include: cars, houses, holidays, close relatives' birthday presents and wedding clothing.

Low involvement will occur as a state of mind when there is very low risk or interest in the product category or it is a highly routinised purchase. Typical products include groceries, postage stamps and office supplies.

Involvement theory also needs to be considered in relation to the *type* and context of purchase scenario as discussed previously.

3.5 The purchase decision-making process

Purchase decision making is very important to understand because it helps marketers to identify how consumers are likely to respond to different elements of the marketing mix throughout different stages of their buying process. Unfortunately many marketers make the mistake of only considering the impact of the promotional aspects of the mix and don't consider other elements such as distribution strategy which is essential for helping to facilitate a purchase.

Global case study

A consumer works their way through a decision to purchase a new luxury pen. They have seen a number of adverts outlining the benefits of the pen, identify that the brand of pen is desirable to them because it refelcts how they would like to be viewed as a professional during meetings however...

...when the consumer visits local stores, the brand is not stocked by any retailers. Undetered, the consumer visits the brand website to find that there is no list of stockists and no facility to purchase directly online.

You should be able to see the key problem here. Only the most highly involved consumer will persevere at this stage and so no matter how much communications budget was dedicated to creating a promotional campaign, in the event very few sales will arise as a result and the brand may be at risk of being tarnished because of the frustration felt by consumers who cannot easily make their purchase.

3.5.1 A simple decision process model

Imagine the following scenario when you are at an airport and feel suddenly hungry. The behaviour or possible thoughts of the consumer at each stage is likely to vary considerably (because of their culture, tastes, purchase opportunities etc). Each **stage of behaviour**, at least at a simple level, can be distinguished as shown in the following table.

Buying process stage	Purchase scenario = Buying a snack at an airport
Need recognition	Feeling hungry but the flight is due to be called to board.
Information search	Look for a vending machine, cafe or shop. Find a shop and cafe.
Evaluation of alternatives	Consider alternatives such as hot, cold – sweet or savoury, decide on sweet, view alternative sweets in nearest shop – a chocolate bar close to the till is chosen for reasons of speed and ease.
Purchase decision	Make a fast decision based on known brands – Mars bar.
Post purchase evaluation	Glad chose Mars bar rather than hot meal, as in the event the flight began boarding very quickly following purchase.

The table below now outlines in detail each of the purchase stages and some issues that marketers need to consider.

Element	Comment
Need/problem recognition	Leads to motivation. (In a more complex model, would be shown to be triggered by psychological, physiological and social factors.) Marketers can identify needs/problems for the consumer, through product positioning.
Pre-purchase/ information search	In a more complex model, would be affected by sources, attitudes, perceptions. Marketers can provide product information, tailored to need.
Evaluation of alternatives	Marketers can make products available for evaluation and provide comparative information about competing products: the important thing, though, is to get the product onto the short-list of options.
The purchase decision	(In a more complex model, would be affected by situational factors: intention is not everything.)
Post-purchase evaluation	Experience 'feeds back' to the beginning of the process, providing positive or negative reinforcement of the purchase decision. If the consumer is dissatisfied, he will be back at the problem recognition stage again. If the consumer is satisfied, the next decision process for the product may be cut short and skip straight to the decision, on the basis of loyalty (or, in a more complex model, expectations and learning constructs).

Such a model provides a useful descriptive framework for marketers, with indications of strategy requirements at each stage. It is still, however, too simple to be predictive, unless a large number of extra social, psychological and cultural variables are added or assumed. This means that the marketer will need to remember the strategies involved and provide the context for different purchase scenarios.

Discussion

Cohort

Think about alternative buying scenarios such as when buying clothes. The way we behave when we buy an outfit to wear to a wedding, work clothes or beachwear differs considerably because there are different levels of importance and risk associated with the need to make a positive purchase. What might be the details which are not included in the simple decision making process model associated with the purchase?

It is often argued that the buying process is more relevant for high involvement products in an extensive problem solving situation (Smith & Taylor, 2007). Routinised response buying (buying a newspaper) may not have easily identified distinct stages because the decision happens so quickly.

3.5.2 The Engel, Blackwell and Miniard model

This model was originally developed in 1968 by Engel, Kollat and Blackwell but has since undergone several revisions. The 1990 version is considered here.

The basis of the Engel-Blackwell-Miniard model takes the simple process of consumer behaviour described above as its basis but then adds more context and detail.

The **influencing variables** are divided into four main categories.

- Stimulus inputs
- Information processing
- Decision process
- Variables influencing the decision process

Step 1

The starting point of the process is the customer's perception of a want or a problem which must be satisfied.

Step 2

This stimulates the start of the next stage, information search, which can be divided into two further stages. First, the customer searches his **internal memory** to ascertain what is known about potential solutions to the problem. If insufficient information is found through this course of action, the customer will begin the process of **external search**. The likelihood of external search is also affected by environmental factors, such as the urgency of the need, and also the individual characteristics of the customer. For example, individuals who are low risk-takers will tend to seek more information before making a decision.

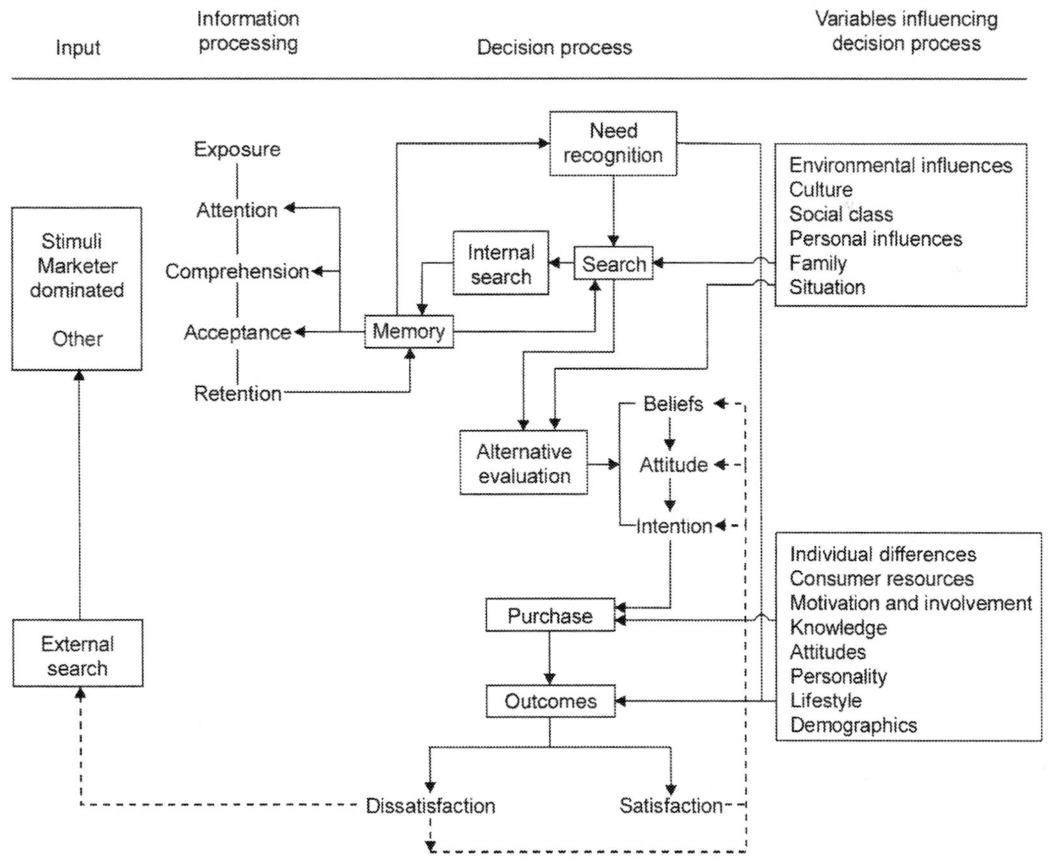

The Engel-Blackwell-Miniard model

Step 3

This search process identifies the various ways in which the problem can be solved. Those various ways are then evaluated. The **alternative evaluation** stage involves comparing the alternative brands against evaluative criteria, which are 'product judging standards that have been stored in the permanent memory'. This evaluation process may lead to changes in beliefs regarding the brands, which, in turn, leads to changes in attitudes and intentions to purchase.

Step 4

The process of **alternative evaluation** leads to an **intention** to make a purchase of the most favourably evaluated brand.

Step 5

This intention will be translated into action unless unforeseen circumstances intervene to postpone or prevent the purchase.

Step 6

Once purchased, the customer will use the product and will continue to evaluate the product by comparing performance against expectations. If the product chosen does not meet expectations, the result is dissatisfaction and this may lead to further search for information about the brand and/or changes in beliefs.

The overall process can, therefore, be seen as a continuous one, especially given the desirability of repeat purchase.

4 Models of buyer behaviour

4.1 The use of models

Modelling is based on the idea that any phenomenon or process can be simplified, by leaving out of the 'picture' any aspects or variables that are not of interest to the modeller, while still portraying something meaningful about the real phenomenon or process. This is particularly true of consumer behaviour models.

Because a model is a simplification of reality based on the modeller's interests, different models may be developed to describe the same phenomenon or process, showing different aspects of the same thing. A range of consumer behaviour models exist (and continue to develop).

4.2 Hierarchy of effects models

We have just looked at the Engel-Blackwell-Miniard model. There are a number of additional models relating to consumer behaviour and the most established fall within the category of Hierarchy of effects. These models generally assume that things happen in a certain order and that earlier effects form necessary conditions in order for later effects to occur (De Pelsmacker, Geuens & Van den Bergh, 2007).

These models stipulate that consumers go through various stages when responding to marketing communications. These stages are cognitive, affective and conative (as discussed in section 2 of this chapter).

Cognitive stage	Consumers engage in thinking processes which lead to awareness and knowledge of the brand.
Affective stage	Emotional responses occur which are associated with the brand.
Conative stage	Undertaking actions eg buying.

The table below is adapted from Barry and Howard (1990) and refers to a variety of hierarchy of effects models.

Model	Cognitive	Affective	Conative
AIDA St Elmo Lewis (1900)	Attention	Interest, desire	Action
AIDAS Sheldon (1911)	Attention	Interest, desire	Action, satisfaction
AIDCAR Kitson (1921)	Attention	Interest, desire, conviction	Action
Lavidge and Steiner (1961)	Awareness, knowledge	Timing, preference, conviction	Purchase
AIETA Rogers (1962)	Awareness	Interest, evaluation	Trial, adaption
ACALTA Robertson (1971)	Awareness, compensation	Attitude, legitimisation	Trial, adoption

4.3 Black box models/stimulus-response

Black box models assume that **observable behaviour** is the only valid object of study. Concentrating as they do on environmental factors, such models in the context of consumer behaviour are **market models** which may be used in market research to identify, for example, the following.

(a) The **decision environment**. Factors external to the individual which influence his buying behaviour, are shown – but the individual is a black box.

(b) The **marketing distribution process**. The 'flow' of products, or information, or influence, is plotted – from producer to salesforce to retail outlet to consumer to other consumers. Competitors can be similarly plotted on the model.

(c) The **buying process** – taking into account only inputs and outputs.

Black box models

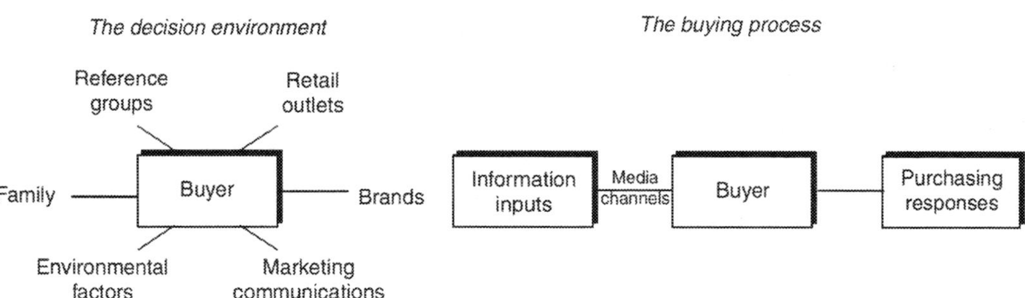

Why are Black box models useful?

(a) They include **observable, quantifiable variables** which are easier to measure and to manipulate.

(b) They concentrate on a **manageable number of relevant input variables**, on which a strategy can be based – without 'analysis paralysis' from speculating about all the possible intervening factors.

(c) **Stimulus variables** such as price, quality, availability, service, or advertising, can be identified by models, and the results of each – in terms of product/brand or supplier choice, quantity and frequency of purchase – set out in a simple, direct way. If a model indicates that decreased price results in increased purchase quantity, the marketer is able to respond accordingly; it does not matter, to an extent, why the phenomenon occurs.

(d) Black boxes are, however, limited to **simple, unambitious functions**. They do not attempt to predict behaviour in a wide range of circumstances, nor to explain behaviour.

Analyse a simple regular purchase that you make – a litre of milk, say, or a chocolate bar – in terms of the black box model depicted above.

4.4 Personal variable models

Personal variable models are simple models of **internal processes** – beliefs, intentions, motives, perceptions etc – without any of the **external, environmental influences**.

Examples of personal variable models

(a) Compensatory or **trade-off model**. When faced with a choice between products composed of certain attributes or benefits in different proportions, a consumer will compromise on his image of the 'ideal' product, and accept less of one attribute in return for more of another. This is a judgement about which combination of attributes offers the highest overall utility, or value.

(b) The **threshold model**. For each perceived attribute of a product, there is a perceived threshold of acceptability, below which the product will be rejected. Each attribute is assessed until the product falls short on one (usually, price) and is discarded.

4.5 Diffusion of innovation model

New products represent a significant opportunity to both **consumers** (who may find a better means of satisfying their needs) and **marketers** (who may find a new source of profit or competitive advantage). The **first buyers** of a product are the **most important**, since there is no established frame of reference within which the product will be considered: no experience, association or reference group adoption to build on.

Definition

An **innovation** is anything new (product, service, practice or idea) from the point of view of:

(a) The organisation ('We've never done this before').
(b) The product ('It hasn't been done quite like this before').
(c) The market ('You may not have seen this before').
(d) The consumer ('I don't think I've seen this before').

Everett Rogers, (1962) definition of innovation is: 'any idea or product perceived by the potential innovator to be new'. This is the essence of the customer-focused approach to innovation.

Diffusion of innovation is the 'macro' process by which the innovation is spread or disseminated from the source to the consuming public.

Adoption is the 'micro' process by which a consumer makes the decision to accept or reject an innovation.

You may have noticed that some new ideas 'catch on' suddenly (eg Crocs shoes and iPods), while others take a long time to gain acceptance (eg web only based banking), and others never get beyond the fringes (eg liquid furniture polish). The **rate and extent of diffusion** of an innovation depend on:

- The **characteristics** of the innovation/new product
- The **channels of communication** used
- The **social system** within which communication takes place
- The **stages of the adoption process** reached by members of the social system

4.5.1 Product characteristics that influence diffusion

(a) **Relative advantage**

Relative advantage is the degree to which potential consumers perceive the product innovation to be better than previous or competing products.

(b) **Compatibility**

Compatibility is the degree to which potential consumers perceive the product innovation to be comparable or consistent with their existing values, attitudes, needs and practices.

(c) **Complexity** is the degree to which potential consumers find the product innovation **difficult to understand or use**.

(d) **Trialability**

Trialability is the degree to which potential consumers can test or sample a product innovation before committing themselves to adopting it.

(e) **Observability**

Observability, or 'communicability' is the degree to which a product innovation's benefits or attributes are visible to the potential consumer – by his own observation, or imagination, or the descriptions of others.

Fashion items, for example, have high 'social visibility' and are more easily disseminated than products for private use, which are 'shared' less with other people.

Activity 4

Whenever you see a product or service which claims to be 'new' in some way, apply the tests of relative advantage, compatibility, complexity, trialability and observability to it.

4.5.2 Rate of diffusion: 'adopter categories'

Diffusion research has indicated that diffusion of an innovation follows a normal distribution (a **bell-shaped curve**) over time, and that consumers can be classified according to the time they take to adopt an innovation, relative to other consumers.

Five adopter categories have been identified (by Rogers (1962) and others). (Note that although they are 'adopter' categories, we are still looking at 'diffusion', not 'adoption' as a process: we are still at the 'macro' level of society and the whole life of the product innovation.)

(a) **Innovators** (2.5% of the population that eventually adopts the product).

Innovators are the first people to adopt an innovation. Their main characteristic is said to be 'venturesomeness'. They are not averse to risk, are eager to try new ideas, are varied and extensive in their social networking.

(b) **Early adopters** (13.5% of the population that eventually adopts the product).

Early adopters are next to adopt the innovation. Their main characteristic may be called 'respectability': they take fewer risks than innovators (watching to see how they get on before themselves adopting the innovation), and are integrated into their social system and culture. They have the highest number of opinion leaders and role models, and so are important in the communication process.

(c) **Early majority** (34% of the population that quickly adopts the product).

The early majority are the first of the general mass of the population to adopt a new idea: just before the 'average' adoption time. Their main characteristic is said to be 'deliberation'. They are

slightly above average in education, age and income, but tend to be followers, seldom holding leadership positions, and relying heavily on information from others.

(d) **Late majority** (34% of the population that eventually adopts the product).

The late majority are the last of the general mass of the population to adopt a new idea: just after the 'average' adoption time. Their main characteristic is said to be 'scepticism'. They are cautious about new ideas and tend to adopt only as a result of economic necessity, or social pressure.

(e) **Laggards** (16% of the population that eventually adopts the product).

Laggards are last to adopt. Their main characteristic is said to be 'traditionalism'. They tend to be oriented to the past and custom, parochial in their social outlook and suspicious of anything new.

Remember that these are **not stages** people go through: they are **'types'** or categories of people. The stages refer to the **diffusion process over time**, and the points in it at which the different categories of people adopt the innovation.

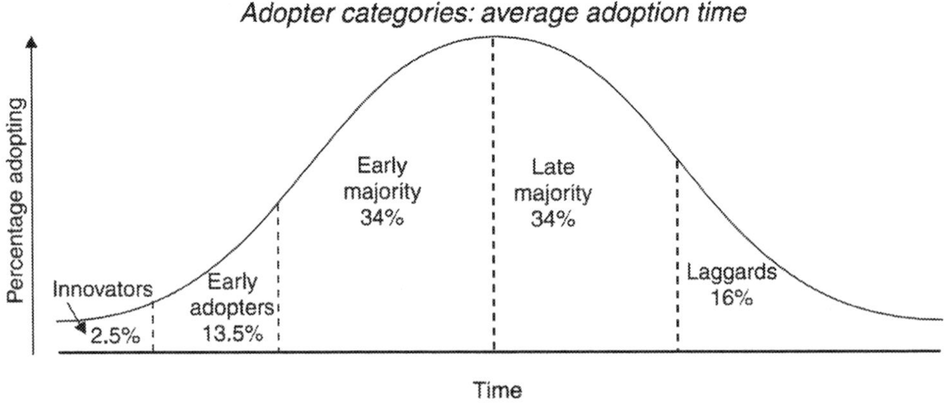

The main problem with using this model is that the categories appear to add up to 100% of the social system – the target market. This is not a reflection of marketers' experience, since some potential consumers do not adopt/purchase at all.

4.5.3 The adoption model

> **Definition**
>
> **Adoption** is the process by which a consumer arrives at a decision to try (or not to try) and – having tried – to continue using (or discontinue using) a new product.

Based on the premise (not universally accepted) that consumers tend to engage in fairly extensive information search and problem-solving in order to reach a purchase decision, a five-stage model of product adoption has been proposed.

(a) **Awareness**. Consumers have been exposed to the product innovation, and are aware that it exists, but their attitudes to it are neutral at this stage.

(b) **Interest**. Consumers become curious about the innovation or aware of its potential to fulfil a need or want of theirs: they become interested in it to the extent of seeking out more information.

(c) **Evaluation**. Consumers use the information they have gathered in order:

(i) To decide whether more information is necessary: delayed decision.

(ii) To establish a positive or negative attitude (favourable or unfavourable evaluation) of the innovation.

(iii) To decide whether to purchase/try or reject the innovation.

The perceived product characteristics discussed earlier will influence this stage.

(d)   **Trial.** Consumers try out the innovation, if possible on a limited, low-risk basis: their experience then provides them with the decisive information to adopt or reject.

(e)   **Adoption** (or rejection). Based on evaluation ('mental trial') and physical trial of the innovation, consumers decide to continue to use it on a full committed basis – or to reject it.

Some researchers suggest that following trial, there are **two intermediate stages** (direct product experience and product evaluation: confirmation) before rejection or adoption.

4.5.4 Value of the adoption model

The adoption model offers a useful framework for marketers.

(a)   To concentrate on **relevant aspects of the consumers' state of mind** at each stage. The most important goal, initially, will be to get attention, in competition with other stimuli. The next will be to arouse interest by suggesting motivations. The next will be giving information relevant to evaluation, then making the product available for trial etc

(b)   To study **which information media are most effective at each stage**. Awareness is best served by mass media, but in later stages, especially evaluation, mass media influence declines in favour of more personal sources: salespeople, opinion leaders and so on.

5   Decision making units

In practice, many different people are involved in taking the buying decision. We hinted at the various different roles – the consumer is not necessarily the person who pays for the good.

> **Definition**
>
> **Decision-making unit** – all people influencing a buying decision – sometimes this may constitute just one person.

Role	Comment
Initiator	Suggests the idea of buying a particular product
Influencer	Provides information about a product/service to other members of the DMU.
Gatekeeper	Controls the flow of information about a product/service into the family (for example by 'giving the gist' of a consumer article, or selecting the advertising message that is relayed to the family).
Decider	Has the power to determine whether to purchase a specific product/service.
Buyer	Makes the purchase of the product/service.

5.1 The family DMU

Within a family group as a decision-making unit various members may occupy these roles. Let us consider the examples of the purchase of a child's toy and a family car.

Toy		Car	
Initiator	Child, parents	**Initiator**	Main earner, salesperson, family
Influencer	Parents	**Influencer**	
	Relatives		Family
	Friends		Friends/colleagues
	Salesperson		Salesperson
Decider	Parents, child	**Decider**	Parents
Buyer	Parents	**Buyer**	Parents
User	Child (and possibly parents!)	**User**	Parents (and perhaps older children)

Marketers will be interested in reaching each of these roles: to make the product attractive to the initiator (if it is not a product automatically bought), **influencer** and **decider** in particular. They would need to make the product readily **available** for the **purchaser**, and satisfying enough for the user to initiate future repurchase.

These roles – and the number and nature of the individuals who adopt them – will **vary from product to product, and from family to family**.

If the **buyer** of a product is a **different** person from the **decision-maker**, the marketer would need to **redirect promotional effort** from packaging and point of sale (attracting the buyer in-store) to advertising and promotional activity outside the store (where the decision-maker would be reached). If powerful influencers can be identified, they can be targeted as a way of getting to decision-makers.

Media which are designed for **sharing by the whole family** unit like television, Sunday newspapers, or cinema (when 'family' films are showing) offer marketers the opportunity of reaching influencers, decision-makers, buyers and users at the same time, facilitating discussion and influence within the decision-making unit.

5.2 The influence of gender

The pattern of decision-making roles has also **varied in society generally**, over time, particularly in relation to **gender roles**. The relative influence of male and female partners is of particular interest to consumer researchers, who commonly classify decisions as being male-dominated, female-dominated, joint, equal or 'syncratic', or unilateral or autocratic.

The relative influence of male and female partners, and the extent to which decisions are shared, may vary according to a variety of factors.

(a) The **family's attitudes to gender roles**.

(b) The **surrounding culture's attitudes to gender roles**.

(c) The **product or service**. Car purchase is a particularly interesting example: several decades ago, it was strongly male-dominated, while now, although still mainly male-dominated for the purchase of the main family car, the female car-buyer forms a rapidly expanding market segment (for second cars, cars for single and/or working women etc). The reverse trend is true for products like food and household items.

As the couple's education increases, more decisions are likely to be made together.

5.3 The influence of children

Children also have (or attempt to have) an influence on purchase decisions, not only those which are **relevant** to their particular wants ('I want toy X', 'I want chocolate bar Y') but also those which – **however irrelevant to them personally** – they relate to attractive messages they have seen and heard on television, or at friends' houses (a brand of dishwashing liquid marketed as having lots of soft bubbles, a car that a friend's dad has and so on).

5.4 The DMU within organisations

All **organisational buying decisions** are **made by individuals or groups of individuals**, each of whom is subjected to the same types of influences as they are in making their own buying decisions. However, there are some real **differences** between personal and organisational buying decision processes.

(a)   Organisations buy because the products/services are **needed to meet wider objectives**: for instance, also to help them to meet their customers' needs more closely. (However, as individuals also do not buy products and services for themselves but for the benefits they convey, it could be argued that both are forms of derived demand.)

(b)   A **number of individuals are involved** in the typical organisational buying decision. Again, a personal buying decision may involve several family members, for example, and so may not be very different in some cases.

(c)   The **decision process may take longer** for corporate decisions. The use of feasibility studies, for example, may prolong the decision process. Tendering processes in government buying also have this effect.

(d)   Organisations are **more likely to buy a complex total offering** which can involve a high level of technical support, staff training, delivery scheduling, finance arrangements and so on.

(e)   Organisations are **more likely to employ experts** in the process.

There are many examples of different types of organisational buying behaviour ranging from simple reordering (for example, of stationery supplies) to complex purchasing decisions (such as that of a consortium to build a motorway with bridges and tunnels, for example).

One system of categorisation for corporate buying decisions, discussed by Howard and Sheth (1969), is based on the **complexity of organisational behaviour**. They identify three types:

Routinised buyer behaviour	This category is the habitual type, where the buyer knows what is offered and is buying items which are frequently purchased. It is likely that the buyer has well-developed supplier preferences and any deviation in habitual behaviour is likely to be influenced by price and availability considerations.
Limited problem-solving	This category is relevant to a new or unfamiliar product/service purchase where the suppliers are nevertheless known and the product is in a familiar class of products, for instance a new model of car in a company fleet.
Extensive problem-solving	This category relates to the purchase of unfamiliar products from unfamiliar suppliers. The process can take much time and effort and involve the need to develop criteria with which to judge the purchase. For example, the construction and refurbishment of new offices where previously buildings were looked after by managing agents.

5.5 The DMU in a business-to-business context

The **DMU is a particularly useful concept** in marketing industrial or government goods and services where the customer is a **business or other organisation**. The marketing manager then needs to know **who** in each organisation makes the effective buying decisions – this might be one person, or a group of people – and the DMU might act with formal authority or as an informal group reaching a joint decision. Many large organisations employ **purchasers** – but the autonomy of the buyers will vary from situation to situation.

Purchasing decisions may be influenced by several people in the consuming organisation.

(a)   **Employees or managers** in operational departments might make recommendations about what type of supplies should be purchased.

(b)   A **junior purchasing manager** might decide what he would like to buy, but **submit his recommendation** to a superior for approval.

(c)   In large organisations, there will be **several purchasing managers**, who might work independently, but might also work closely together, either formally or informally.

(d)   Technical specifications for component purchases might be provided by **engineers or other technical staff**.

(e)   **Accountants** might set a limit on the price the organisation will pay.

(f)   Large items of purchase might require approval from the **board of directors**.

The relationships between **members of the DMU** are also important.

The **user** may have influence on the technical characteristics of the equipment (and hence the cost) and on reliability and performance criteria, and so on.

6 Buyer motives

6.1 Theories of human motivation

Definition

Motivation has been defined as a psychological force that energises, activates and directs behaviour towards goals.

Motivation arises from perceived needs. These needs can be of two main types – biogenic and psychogenic.

(a)   Biogenic needs arise from physiological states of tension such as hunger, thirst and discomfort

(b)   Psychogenic needs arise from psychological states of tension such as the need for recognition, esteem or belonging.

Most needs are not intense enough to motivate an individual to act immediately, but when aroused to a sufficient level of intensity the individual will be motivated to act in order to reduce the perceived tension. We prioritise the effort we make towards satisfying our needs, generally giving more time and effort to those needs that have higher costs and benefits, and that are more important, interesting and relevant to us.

Maslow's theory of motivation seeks to explain why people are driven by particular needs at particular times. Maslow (1954) argues that human needs are arranged in a hierarchy comprising, in their order of importance: physiological needs, safety needs, social needs, esteem needs and self-actualisation needs. Maslow states that a person will attempt to satisfy the most important need first. When that need is satisfied it ceases to be a motivator and the person will attempt to satisfy the next most important need. For example, if you are hungry (a physiological need) you will venture out from your desk to get a sandwich.

Herzberg (1966) developed a 'two factor theory' of motivation that distinguishes between **factors that cause dissatisfaction and factors that cause satisfaction**. The task for the marketer is, therefore, to avoid 'dissatisfiers' such as, for example, poor after-sales service, as these things will not sell the product but may well unsell it. In addition the marketer should identify the major satisfiers or motivators of purchase and make sure that they are supplied to the customer.

6.2 The consumer's motivation mix

Customer behaviour is determined by **economic, psychological, sociological** and **cultural** considerations. The reasons for buying a product may vary from person to person, or product to product, or the reasons

may be the same, but the weighting given to each reason in the mind of the customer may vary. These reasons make up the motivation mix of the customer.

The **motivation mix** leads a customer to choose:

- The type of goods or services they want to buy (food or a new haircut, say)
- The brand
- The quality
- The quantity
- In what place
- From whom
- At what price
- By what method of payment (cash, cheque, credit card)
- The timing of their purchase

The motivation mix of a customer for **consumer goods** will be different from that of a buyer of **industrial goods**.

(a) The **domestic buyer** might buy on impulse, attracted by the branding, packaging or display of an article, as well as to meet the family's needs within the shopping budget. The homemaker has a variety of demands to satisfy, within the limits of the family budget.

(b) An **industrial buyer** might be expected to give more emphasis to rational motives for purchasing – a clear need for the article, its price and quality, delivery dates and after-sales service.

Motives sit between **needs and action**. Motives are derived from needs in that a need motivates a person to take action. The difference between needs and motives are as follows.

(a) Motives **activate behaviour**. Being thirsty (need) causes (motivates) us to buy a drink (action). If the need is sufficiently intense, we are motivated to act.

(b) Motives are **directional**. Needs are general but motives specify behavioural action. A general need to belong may lead to a specific motivation to join a rugby club.

(c) Motives serve to **reduce tension**. If we are too cold, we are motivated to reduce the tension in our bodies that this causes by seeking a source of warmth.

Although we know that motives arise from needs and can lead to purchase actions, motives do not tell us **how** consumers choose from the options available to satisfy needs. Other influences are clearly at work.

6.3 Psychological influences on buyer behaviour

6.3.1 Attitudes

Some writers have suggested that a buyer's attitude towards a product is an important element in the buying decision. However, there is no empirical evidence to link a favourable attitude directly with consumer purchases.

6.3.2 Loyalty

A consumer may demonstrate a loyalty towards a particular company's goods or to a brand, or even to a particular shop or retail chain. It is not clear, however, that marketing efforts can be successful in trying to increase customer loyalty. Although it is an important phenomenon, it may well be outside the sphere of management influence or control.

6.3.3 Personality

Inevitably, a consumer's buying behaviour will be influenced by their personality. For example, the latest computer might be purchased by a customer whose personality traits include self confidence, dominance and autonomy. Coffee producers, on the other hand, have discovered that heavy coffee drinkers tend to have sociability as a principal personality trait.

6.4 Attitudes and behaviour

An attitude is a relatively consistent, learned predisposition to behave in a certain way in response to a given object.

(a) Relative consistency. Attitudes are not permanent; they can be changed. However, they tend to be reasonably well established, and so lead to behaviour with a reasonable degree of predictability. (If someone prefers coffee to tea, they are likely to choose coffee over tea fairly consistently.)

(b) Learned. Attitudes are 'learned' or conditioned, formulated with experience as a result of learning factors such as motivation, association and reinforcement.

(c) Predisposition. Attitudes are a predisposition to behaviour; they do not imply that a given behaviour will necessarily follow.

(d) An object. Attitudes relate to some aspect of the individual's environment – it may be a thing, person, event, concept or whatever.

> **Definition**
>
> An **attitude** is a learned predisposition to behave in a consistently favourable or unfavourable way with respect to a given object (Schiffman and Kanuk, 2004).

There is no automatic, direct relationship between attitude and behaviour. Insofar as attitudes and behaviour are inter-related, the process may be portrayed as a tendency for an individual to behave in a given way: not a prediction of how he will behave.

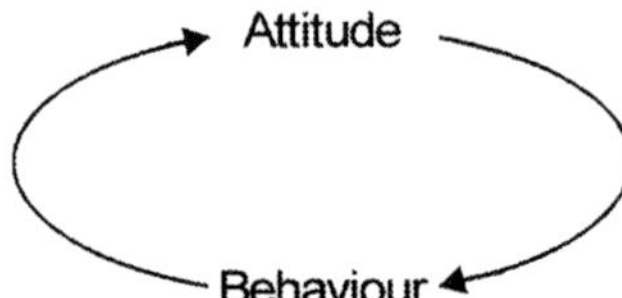

Situational factors affect our behaviour; attitudes are not the sole variable involved.

(a) An individual's behaviour may be influenced by the situation, to contradict his attitudes. For example, you may have a strong liking for organically-grown produce (you believe it is 'right' to buy such produce, and you intend to buy it) – but it is more expensive in the supermarket, so for economic reasons you do not buy it every week.

(b) An individual's attitudes may also be influenced by the situation. People who generally avoid chocolate for health and diet reasons may feel differently in situations where they are physically exhausted and need a boost in blood sugar, or where they are in a festive mood and sharing a bar/packet/cake with friends – especially if told that 'A Mars a day helps you work, rest and play'!

6.4.1 Attitude formation

Attitude theories are based on the premise that individuals seek cognitive consistency.

(a) Consistency between the attitudes they hold.
(b) Consistency between their attitudes/perceptions and their experience of reality.
(c) Consistency between their behaviour and their self-image.

When discrepancies or inconsistencies occur (and are perceived) the individual experiences tension. In order to alleviate that tension, the individual has to change one or more of the factors creating the inconsistency: to change one of their conflicting attitudes to 'fit' the other, to change their attitude to 'fit' their behaviour, or to change their behaviour to 'fit' their attitude.

Balance theory is simple, which is useful when it comes to application. However, it has been criticised as too simple to adequately portray the complexity of attitude systems and conflicts, particularly since it uses

only the dimensions 'positive' and 'negative' without taking into account the strength of positive or negative feeling, which would allow us to predict which way people would choose to resolve an unbalanced situation (Heider, 1958).

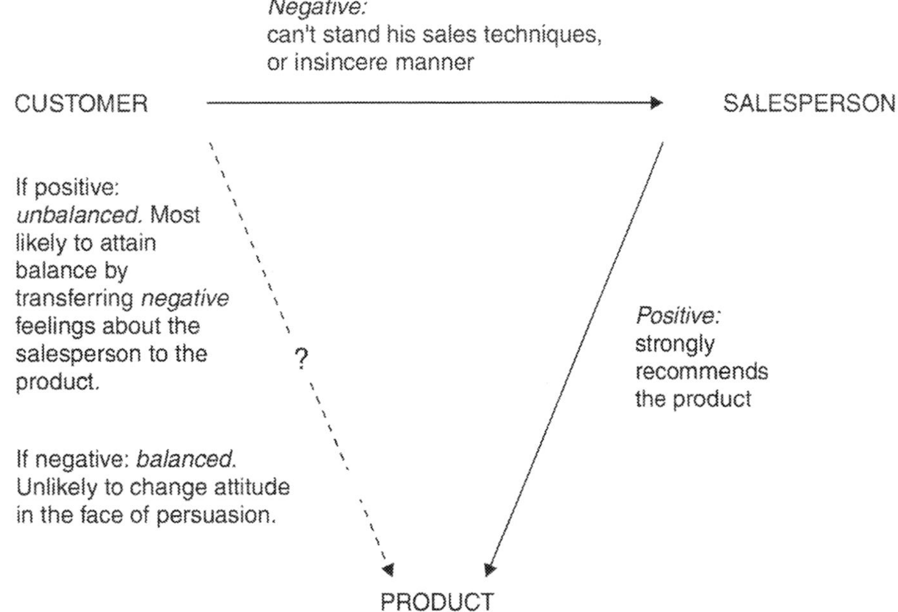

An application of balance theory

Balance theory – considers relations among elements which a person might perceive as belonging together, and people's tendency to change relations among elements in order to make them balance.

6.4.2 Cognitive dissonance theory: Festinger (1957)

Cognitive dissonance is the discomfort experienced by an individual when they receive new information which appears to contradict a belief or attitude they hold. The theory is particularly concerned with decision-making.

(a)	In making a decision – for example, to buy a product – individuals frequently have to weigh up positive and negative factors and compromise on that basis. Post-purchase dissonance occurs when somebody regrets or feels anxious about a choice they have made.

(b)	They may also find themselves acting in a way that appears to contradict a belief or attitude they hold. They have chosen a foreign-made car because of a special financing deal or promotional campaign, when they had previously always believed in 'buying British'.

The individual may experience psychological discomfort as a result.

Marketers can use dissonance theory in several ways.

(a)	To persuade people to try or buy first and develop a positive attitude afterwards (eg money back if not satisfied).

(b)	To reinforce purchase decisions. The consumers will only be able to convince themselves they like the product if you tell them in your advertising they've done the right thing!

(c)	Be careful to avoid disappointing your customers.

Post purchase behaviour is as important to manage as pre-purchase behaviour. Remember that word of mouth communications and reviews are increasingly important in the age of digital media.

It is frequently stated that most car adverts are actually designed to appeal to consumers who have already purchased their vehicle and help them satisfy themselves that they have made a worthwhile purchase.

6.5 Socio-psychological influences

According to Williams (1981), socialisation is 'the process by which the individual learns the social expectations, goals, beliefs, values and attitudes that enable him to exist in society'. In other words, socialisation is the process by which an individual acquires sufficient knowledge of a society and its ways to be able to function and participate in it.

The learning of gender-related, consumer and occupational roles is part of the socialisation process: what it means to 'be' or 'behave like' a girl or boy, what money is for, what 'buying' is, what 'work' is and what sorts of work different 'sorts' of people do.

We can therefore identify a number of agencies who are instrumental in the socialisation of individuals.

(a)   The family is probably the most enduring and extensive source of influence, because it is here that the dependent child learns from their parents and siblings. The family has a particularly strong influence on the child's perception of appropriate roles and on their self-esteem, which in turn affects aspirations.

(b)   School is an important source of values, particularly as they relate to other people.

(c)   Peer groups exert influence on social groups which can control a person's social and emotional satisfactions, say by appointing a person unelected leader or by 'sending him to Coventry'. As a person grows, peer groups become more of an influence, eventually superseding the family.

(d)   The mass media is extremely pervasive and has a profound effect on all consumers.

'Consumer socialisation' is the process by which children acquire the skills, knowledge and attitudes that enable them to function in society as consumers.

(a)   Children observe the consumption behaviour of their parents or older siblings (while pre-adolescents), or their peers (once adolescents and teenagers), and model their own behaviour accordingly.

(b)   Parents use consumption-related events to socialise children generally: promises of gifts or shopping expeditions are used as incentives to behave in a desired way; withholding of money for self-directed purchases is threatened as a deterrent to undesirable behaviour.

(c)   Children are socialised into attitudes, values and motivations which are indirectly related to consumption: products or particular brands are means of satisfying socialised needs and wants. Socialisation creates consumer motivations.

6.5.1 Reference groups

Definition

A **reference group** is any person or group that serves as a point of comparison (or reference) for an individual in forming either general to specific values, attitudes, or a specific guide for behaviour.

A reference group is an actual or imaginary individual or group perceived as having significant relevance upon an individual's evaluations, aspirations or behaviour.

Groups can serve as a benchmark for:

(a)   General behavioural norms.
(b)   Specific attitudes or behaviour – such as 'fashionable' product purchases.

An individual may be influenced by a non-contractual or secondary group, as well as a primary one with which they are intimately in contact. Note, too, that the individual does not need to be a member of a group in order to measure his behaviour by it.

(a)   An aspirational group can impel an individual to act as it does, or to wear the same 'badges', in order to feel closer to attaining actual membership.

(b)   A dissociative group can impel an individual to disown any behaviour or object associated with it.

Reference groups influence a buying decision by making the individual aware of a product or brand, allowing the individual to compare his attitude with that of the group, encouraging the individual to adopt an attitude consistent with the group, and then reinforcing and legitimising the individual's decision to conform.

Global case study

Group membership has entered cyberspace as 'netizens' around the world rapidly form virtual communities. Members are linked to one another via their computer modems, and all of their interactions are digital. This electronic anonymity opens up exciting new opportunities for many, especially those who have difficulty interacting in face-to-face settings (eg the Internet has made a dramatic difference to the lives of many disabled people who can now interact with others around the world without having to leave home). New technologies allow people to chat about their mutual interests, to help one another with enquiries and suggestions, and to obtain suggestions for new products and services. Websites such as FaceBook and uTube have become major social networking platforms and have grown to become a powerful media option for advertisers.

It has been suggested that we adopt different reference groups for different areas of our lives and consumer choices. Miles carried out a study of adolescent girls and found that the peer group was most influential in, for example, the choice of clothes and books, while in matters such as the choice of boyfriends, their parents' opinions were more valued. Youth, however, is the stage at which we are most personally insecure, and at which the reference group has the most power: consider the market for training shoes or music, for example.

6.5.2 Roles

Definition

A 'role' is the sum or 'system' of expectations which other people have of an individual in a particular situation or relationship. Role theory is concerned with the roles that individuals act out in their lives, and how the assumption of various roles affects their attitudes to other people.

An example may help to explain what is meant by 'roles'. An individual may consider himself to be a father and husband, a good neighbour and an active member of the local community, a supporter of his sports club, an amateur golfer, a conscientious church-goer, a man of certain political views, a professional and a marketer.

The other individuals who relate to him when he is in a particular role are called his role set. At work, he will have one role set made up of colleagues, superiors and subordinates, and any other contacts in the course of business; at home, he will have another role set consisting of family members.

Roles are shaped by:

(a)   The expectations of other people as to how a person in a given role should or usually does behave.

(b)   Norms – the customs and informal 'rules' of behaviour which society has formulated.

We learn, growing up in society and experiencing different role sets, what is expected of us, and what the 'rules' are in given situations.

(a)   We learn, for example, the 'appropriate' behaviour for our gender

(b)   When we start a new job – or first become a student – we are confronted with a whole new set of expectations and norms, and have to 'learn the ropes'

Individuals give expression to the role they are playing at any particular time by giving role signs. One example of role signs is clothing or uniform. A white coat indicates that an individual is performing his role as a hospital doctor, a certain shirt indicates that an individual is acting in the role of a football team supporter, and a school or college scarf is a sign that a person is in the role of student.

Note that role signs can be embodied as particular products. Many products are purchased because they reflect or reinforce social roles. Clothing and accessories are bought as symbols of role, for example, people buy gifts for each other which reinforce the nature of their role relationship (intimate and familial or formal and professional).

The nature of roles is not dependent on the particular people who fill them: people fit into roles. Roles are therefore valid units of analysis, in themselves. Knowledge of roles can:

(a)   Enable marketers to predict the kind of role signs people will want to buy, for a wide range of roles.

(b)   Suggest role models – ideal figures who embody the highest expectations of particular roles – who can be associated with a product (through licensing, advertising or promotions) in order to appeal to the aspirations of people in those roles.

(c)   Offer the basis of market segmentation and product positioning. Categories such as executives, or young mothers, carry with them a range of role expectations, norms and signs which can be appealed to or catered for.

Since roles depend on learning and perception, and since an individual occupies multiple roles, there are situations in which problems occur.

(a)   Role ambiguity is a term which describes a situation when the individual is not sure what his role is, or when some members of his role set are not clear what his role is.

Global case study

Role ambiguity is perhaps inherent in the way certain complex financial services, such as pension plans, have been sold. Is the person that sells the pension a 'financial adviser' or 'an insurance salesperson'?

The difference is important – a financial adviser is required to offer 'best advice' to the customer but, at the same time, may be required to meet sales targets.

Role ambiguity might sometimes emerge as 'conflict of interest'.

(b)   Role conflict occurs when an individual, acting in several roles at the same time, finds that the roles are incompatible. A businessman who receives a telephone call from his wife, who wants him to leave work and go home, will experience conflict in his roles as businessman and family man. Similar conflict might be experienced by a working woman, who must reconcile her roles as a businesswoman and a mother. A trade union member may be reluctant to obey a strike call, because he disapproves of its reasons.

6.5.3 Secondary groups: the social environment

If we are to understand our customers we need to have some idea of how far the individual's cognitive processes and behaviour, including his/her decision whether to become our customer, depend on his/her perception of and interaction with other people. In this section we will cover two particular ways in which we see ourselves, and behave, in relation to others.

Some roles are perceived as superior or inferior in relation to others. The roles of employer and employee, or parent and child, have this connotation in traditional Western societies. One role may be superior to another because it is perceived that a person in that role has power or authority.

A combination of such factors – power, wealth, expertise, position – determines 'where' a person is in relation to other people, or in the framework of society as a whole. This relative position of superiority or inferiority is called status.

In social class research, status is frequently thought of as the relative ranking of members of each social class in terms of specific status factors. For example, relative wealth (amount of economic assets), power (the degree of personal choice or influence over others) and prestige (the degree of recognition received from others) are three status factors frequently used when estimating social class (Schiffman and Kanuk, 2004)

Status may be:

(a)   Ascribed, or attributed to someone by society, on the basis of perceived factors such as their sex or age, intelligence or race. Improved status comes automatically with seniority in some cultures – such as the Japanese – while the male has traditionally had higher status than the female in Western cultures.

(b)   Achieved by deliberate effort on the part of the individual. Professional and occupational status are achieved.

Status may be achieved by acquiring appropriate roles and behaviours.

Activity 5

It is suggested that people play different roles and that their consumption behaviours may differ depending on the particular role they are playing. Do you agree or disagree with this perspective and can you give examples to illustrate?

Some cultures are 'status-conscious', while others are not.

Marketing, as well as management, will need to take into account the extent of status-awareness. Low-status individuals, in a status-conscious society, may aspire to achieve higher status. Products can be positioned accordingly. High status individuals may wish to be 'congratulated', or to emphasise the exclusivity and power of their position in society. Premium quality (and price) products frequently appeal to this sense. Status symbols are products that are purchased and displayed to signal membership of a desirable social class.

6.5.4 Social class: stratification of society

In practice, class strata or divisions are commonly derived from the specific demographic factors of:

- Wealth/income – economic resources
- Educational attainment
- Occupational status

From a marketer's point of view, it is also possible to infer shared values, attitudes and behaviour within a social class, as distinct from those of a higher or lower class. Some research has been able to relate consumption behaviour to class standing. This makes social class an attractive proposition for market segmentation.

6.5.5 Class distinction and mobility

Social class tends to perpetuate itself where people are highly class-conscious – they do not tend to move between classes, or indeed to interact socially with members of other classes. Sociologists note that some societies have a very rigid class system, while others do not.

Social mobility has several implications for marketers.

People aspire to upward mobility, given a reasonable expectation of success, and will exhibit purchase and consumption patterns suitable to the class to which they aspire. (Consider the boom in products for 'yuppies' – by definition 'young upwardly mobile' people in the late 80s – and in demand for owned homes, foreign holidays, private education and health clubs, all of which used to be considered 'exclusive' to the upper middle class.)

The 'lower' classes will become a less significant proportion of the population (and therefore target market), as people move up out of them – assuming that the country's economy can support a widespread increase in per capita income.

People are afraid, in a society of increasing inequality, of downward mobility.

6.5.6 Education and achievement

Throughout the world, education and levels of achievement are highlighted as a means of gaining social status. In some countries academic qualifications are valued highly while in others professional achievements are revered. In the case of this qualification, you have the benefit of both!

6.5.7 Buying patterns

Demography and the class structure are relevant in that they can be both behavioural determinants and inhibitors of household buying behaviour.

(a)   Behavioural determinants encourage people to buy a product or service. The individual's personality, culture, social class, and the importance of the purchase decision (eg a necessity such as food or water, or a luxury) can predispose a person to purchase something.

(b)   Inhibitors are factors, such as the individual's income, which will make the person less likely to purchase something.

Socio-economic status can be related to buying patterns in a number of ways, both in the amount people have to spend and what they spend it on. It affects both the quantity of goods and services supplied, and the proportion of their income that households spend on goods and services.

6.5.8 Secondary groups: cultural issues

Culture is a concept crucial to the understanding of buyer behaviour and can be thought of as the collective memory of a society.

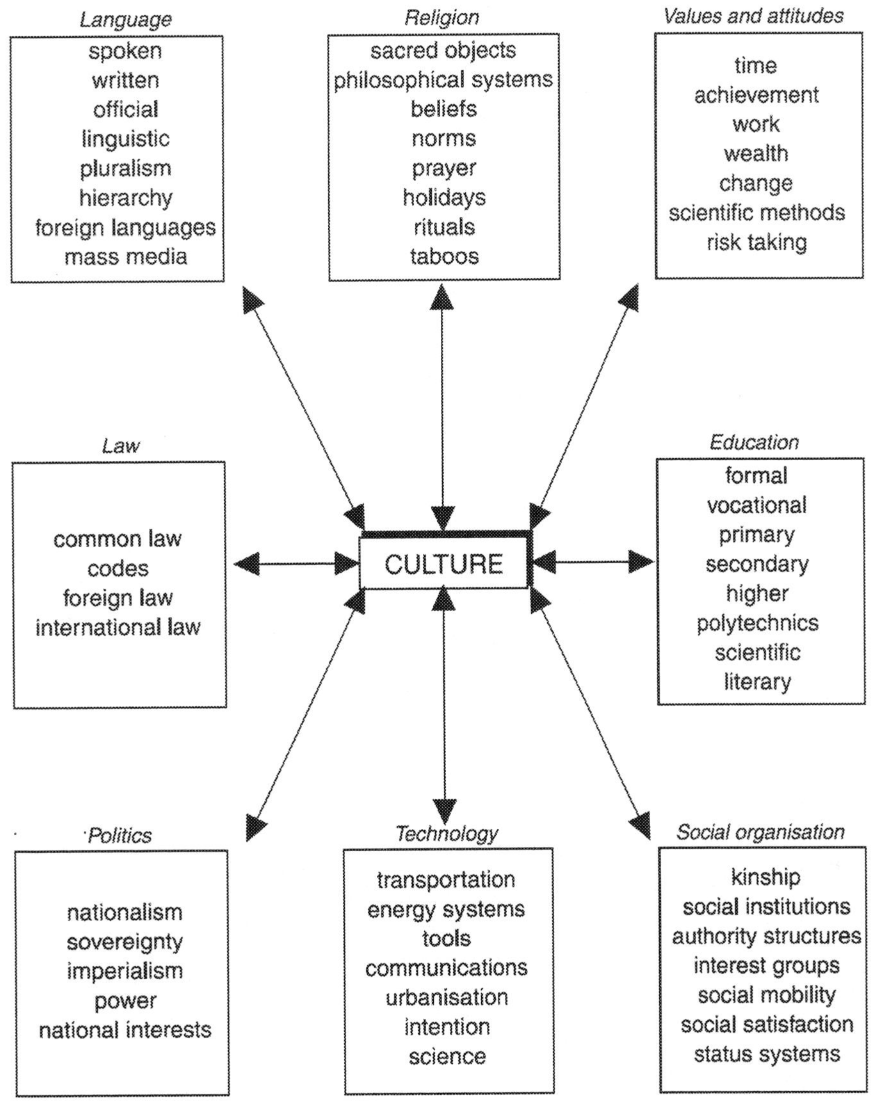

Culture embraces the following aspects of social life.

(a)   Beliefs are perceived states of knowing. We feel that we know about 'things', on the basis of objective and subjective information.

(b)   Values are the comparatively few key beliefs which are:

   (i)   Relatively enduring

   (ii)   Relatively general – not tied to specific objects

   (iii)   Fairly widely accepted as a guide to culturally appropriate behaviour – and therefore as a 'standard' of desirable and undesirable beliefs, attitudes and behaviour

(c)   Customs are modes of behaviour which represent culturally approved ways of responding to given situations: usual and acceptable ways of behaving.

(d)   A ritual is a type of activity which takes on symbolic meaning, consisting of a fixed sequence of behaviour repeated over time. Ritualised behaviour tends to be public, elaborate, formal and ceremonial – such as marriage ceremonies.

(e)   The different languages of different cultures is an obvious means of distinguishing large groups of people.

(f)   Symbols are an important aspect of language and culture. The symbolic nature of human language sets it apart from animal communication. Each symbol may carry a number of different meanings and associations for different people, and some of these meanings are learned as part of a society's culture. The advertiser using slang words or pictorial images must take care that they are valid for the people he wants to reach – and up-to-date.

(g)   Culture embraces all the physical 'tools' or artefacts used by people for their physical and psychological well-being. In our modern society, the technology we use has a very great impact on the way we live our lives, and new technologies accelerate the rate of social change.

      (i)   Appearance and dress
      (ii)   Food and eating habits
      (iii)   Gender roles
      (iv)   Mental processing and learning styles
      (v)   Time and time consciousness

6.5.9 Characteristics of culture

Having examined the areas included in the definition of culture, we can draw together some of the underlying characteristics of culture itself.

(a)   Social. Culture exists to satisfy the needs of people. 'It offers order, direction and guidance in all phases of human problem solving, by providing 'tried and true' methods of satisfying physiological, personal and social needs'.

(b)   Learned. Cultural norms and values are taught or 'transferred' to each new member of society, formally or informally, by socialisation. This occurs in institutions (the family, school and church) and through on-going social interaction and mass media exposure in adulthood.

(c)   Shared. A belief or practice must be common to a significant proportion of a society or group before it can be defined as a cultural characteristic.

(d)   Cumulative. Culture is 'handed down' to each new generation, and while new situations teach new responses, there is a strong traditional/historical element to many aspects of culture.

(e)   Adaptive. Culture must be adaptive, or evolutionary, in order to fulfil its need-satisfying function. Many factors may produce cultural change – slow or fast – in society: eg technological breakthrough, population shifts, exposure to other cultures, gradual changes in values. (Think about male-female roles in the West, or European influences on British lifestyles.)

6.5.10 Micro-culture

Culture is a rather broad concept, embracing whole societies. It is possible to subdivide (and for marketers, further segment) a macro-culture into micro-cultures (or subcultures) which also share certain norms of attitude and behaviour.

Definition

Subculture can be defined as a distinct cultural group that exists as an identifiable segment within a larger, more complex society (Schiffman and Kanuk, 2004).

The members of a specific subculture possess beliefs, values and customs that set them apart from other members of the same society. In addition, they adhere to most of the dominant cultural beliefs, values and behavioural patterns of the large society.

So the main micro-cultures relevant to many countries are defined by the following factors.

(a)   Class (discussed in detail earlier).

(b)   Nationality. Nationality refers to the birthplace of one's ancestors. The UK is home to many nationalities. The key issue for marketers is whether these offer segmentation opportunities.

(c)   Ethnicity. Ethnicity refers to broader divisions. There are identifiable differences in lifestyles and consumer spending patterns among these groups, but it is only relatively recently that attention has been given to reaching and serving ethnic minority market segments, as distinct from mass marketing which would 'also reach' the racial minorities.

(d)   Geography or region. Even in such a small country as the United Kingdom, there are distinct regional differences (eg lifestyle), brought about by the past effects of physical geography (poor communication between communities separated by rivers or mountains, say) and indeed its present effects (socio-economic differences created by suitability of the area for coal-mining, say, or leisure, or urbanisation). There is talk of a 'North/South' divide.

(e)   Religion. Adherents to religious groups tend to be strongly oriented to the norms, beliefs, values, traditions and rituals of their faith. Food customs are strict in religions such as Judaism, Hinduism and Islam. Values and lifestyle are also likely to vary according to the teachings and customs of the religion.

(f)   Age. Age micro-cultures vary according to the period in which individuals were socialised, to an extent, because of the great shifts in social values and customs in this century. (You may have seen specially branded products for 'Baby-boomers': those born between 1946 and 1964 in the post-war birthrate explosion.)

(g)   Gender. We have already discussed gender roles in family buying behaviour, and their gradual erosion – despite which, marketers make frequent appeals to gender-linked stereotypes. The Working Woman and the New Man are perhaps the most important current micro-cultural markets segmented on this basis.

Marketers also need to avoid exaggerating the exclusivity of micro-cultures. Promotional strategies need not target a single micro-cultural membership, since each consumer is simultaneously a member of many micro-cultural segments. (You do not need to sell cornflakes specifically to a ethnic minority, when its members are also Protestants, women, young people, living in the West Midlands – and part of the mainstream UK culture, as far as cornflake consumption is concerned.)

6.5.11 Economic influences

Economic man

Economists have believed that the consumers make rational decisions, that is:

(a)   They are aware of all product alternatives,

(b)   They are capable of correctly ranking the alternatives having assessed their merits and disadvantages, and

(c)   They are able to identify the best alternative.

This probably sounds pretty far-fetched to you and indeed the theory has been roundly criticised on a number of grounds.

(a)   People do not have perfect information.

(b)   People tend to act in accordance with their existing values and goals, and are limited by existing habits, skills and reflexes.

Passive man

(a)   The objections to the economic man model give rise to a second model, the passive man. Such a person is irrational and impulsive, and is therefore entirely prey to the aims and strategies of marketers. The model suggests that the consumer can be manipulated at will.

(b)   It is a simplistic model and ignores the fact that consumers can and usually do seek some information about product alternatives (even if the information is not perfect) and then make a choice which gives them satisfaction, not the marketer (even if the choice is not perfectly satisfying).

Cognitive man

Cognitive man is a thinking problem solver who seeks information on which to base consumption decisions from a range of choices. He is aware of the risks involved in making a choice, and will use various strategies to handle them.

Within this chapter we have worked towards developing your understanding of buyer behaviour.

1 Introduce the concept of buyer behaviour and state why this is important to communications

- Understanding buyer behaviour is important because the **buyer's reaction** to the organisation's marketing strategy has a major impact on the success of the organisation.

- If organisations are truly to implement the marketing concept, they must examine the main influences on **what, where, when and how customers buy**. Only in this way will they be able to devise a marketing mix that satisfies the needs of the customers.

- By gaining a better understanding of the factors influencing their customers and how their customers will respond, organisations will be better able to predict the **effectiveness of their marketing activities**.

- Not all consumers behave in the same way. Decision making and purchase patterns of behaviour vary considerably between individuals and across product categories.

2 Consider how communication works

- Communication is an exchange process.

- The purpose of communication is to differentiate, remind, inform and persuade.

- The process of communication involves a sender encoding a message, transmitting it and a receiver decoding. Feedback is then provided. Noise may interrupt the intended message being decoded in the intended manner.

- Media and people are influences which require careful consideration.

3 Examine the purchase decision making process

- Individuals are important in all buying decisions, either for themselves or on behalf of an organisation. Marketers need to understand their motives, psychology and influences, and how these combine to form a motivation mix.

- Different types of consumer decision-making can be distinguished by the level of time and effort expended.

- Different types of organisational decision-making depend on the nature of the buying decision.

4 Evaluate models of buyer behaviour

- Essentially there are two formats for simple buyer behaviour models, black box models which exclude internal variables and personal variable models which exclude external variables.

- A number of hierarchy of effects models also exist.

- The key elements of the buying process are problem recognition, information search, evaluation of alternatives, the purchase decision and post-purchase evaluation.

- Diffusion of innovation is the term used to describe the process by which an innovation is spread or disseminated from the source to the consuming public.

- The family is of critical importance to individuals as a network of relationships, reference group and social unit. The decision-making unit operates most often in a family context, with different family members adopting different consumption and hence DMU roles.

- Organisational decision-making is based on the same DMU model but is generally more rational yet inert. The interaction of DMU members is influenced by a great many variables. The AMA model can be used to describe this.

- The DMU is a particularly useful concept in marketing industrial or government goods and services where the customer is a business or other organisation.

- The marketing department of a supplier aiming at corporate clients therefore needs to be aware of:

    - How buying decisions are made by the DMU
    - How the DMU is constructed
    - The identities of the most influential figures in the DMU

6 Consider the principles of buyer motives

- Buyer behaviour is determined by economic, psychological, sociological and cultural considerations.

- The term 'culture' encompasses the sum total of learned beliefs, values, customs, rituals, languages, symbols, artefacts and technology of a society or group.

- Different societies vary greatly in most aspects of culture, but all culture is social, learned, shared, cumulative and adaptive.

- Products can be invested with cultural meaning by marketing efforts; that meaning is transferred to a buyer who uses the product in a cultural context and hence imbues it with further cultural meaning. This is called transfer of cultural meaning.

- Micro-cultures within a culture are usually structured on the lines of class, nationality, ethnicity, geography, religion, age and gender.

Now go to Session 1 on the VLE and complete the associated activities.

1   The answer to this activity will depend on your own organisation. If you are having problems, try looking at the communication item and simply itemising points next to the diagram. Think through the following GMN example beforehand if it helps.

GMN hosted a conference in London in 2009 about Chindia Rising. To advertise the event, GMN (among other tactics) sent their network partners a pdf e-flyer.

So far we can deduce the following:

Sender = GMN

Receiver = business partners

You will see annotations referring to this scenario on the following diagram:

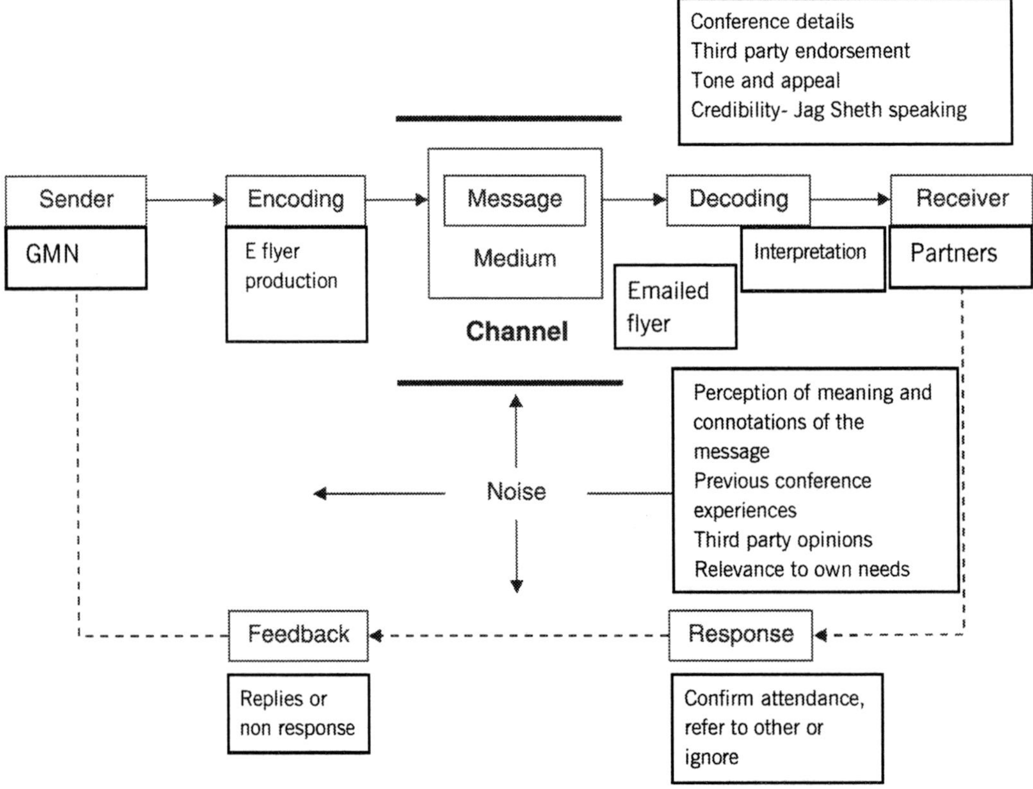

2   This will depend on which products or services you are thinking about. You could consider medical supplies, products aimed at children, office stationery and catering supplies.

3   The result of, or answer to, this activity will depend on the item you have chosen.

4   Sometimes new products are simply 'improved' versions. Think about whether this impacted on your answer.

5   The answer to this activity will depend on the roles you choose.

Barry, T.E. and Howard, D.J., (1990). A review and critique of the Hierarchy of Effects in advertising. *International Journal of Advertising*, 9, pp. 121-35.

De Pelsmacker, P., Geuens, M. and Van den Bergh, J., (2006) *Marketing Communications.* 3rd edition. Harlow: Pearson.

Festinger, L., (1957). *A Theory of Cognitive Dissonance*, Stamford: Stamford University Press.

Fill, C., (2009). *Marketing Communications: Interactivity, Communities and Content.* 5th edition. Harlow: FT Prentice Hall.

Heider, F., (1958). *The Psychology of Interpersonal Relations*, London: Wiley.

Herzberg, F., (1968). *Work and the Nature of Man*. Cleveland: World.

Howard, J.A. and Sheth, J. (1969). *The theory of buyer behaviour*, New York: John Wiley and Sons.

Kapferer, J.N. and Laurant, G., (1985). Consumer involvement profiles: a new practical approach to consumer involvement. *Journal of Advertising Research*, 25 (6) pp. 48-56.

Kotler, P. et al, (1999). *Marketing: An Introduction*. Sydney: Prentice Hall Australia.

Maslow, A., (1954). *Motivation and Personality*. New York: Harper and Row.

Pickton, D. and Broderick, A., (2004). *Integrated Marketing Communications*. 2nd edition. Harlow: Prentice Hall.

Rogers, E., (1962). *Diffusion of Innovation*. New York: Free Press

Schramm, W., (1955). *How Communication Works*. In *The Process and Effects of Mass Communication*. (ed. Schramm, W.) Urbana: University of Illionois Press.

Schiffman, L.G. and Kanuk, L.L., (2004). *Consumer Behaviour*. 8th edition. Oxford: Pearson Prentice Hall.

Smith, P.R. & Taylor, J., (2004). *Marketing Communications: an Integrated Approach*. 4th edition. London: Kogan Page.

Tata, (2009). Tata Motors delivers first Tata Nano in the country in Mumbai.
[Online Press Release] available at:
http://www.tatamotors.com/our_world/press_releases.php?ID=457&action=Pull [Accessed 19.7.09].

Williams, K. C., (1981). *Behavioural Aspects of Marketing*. Oxford: Heinemann Professional Publishing.

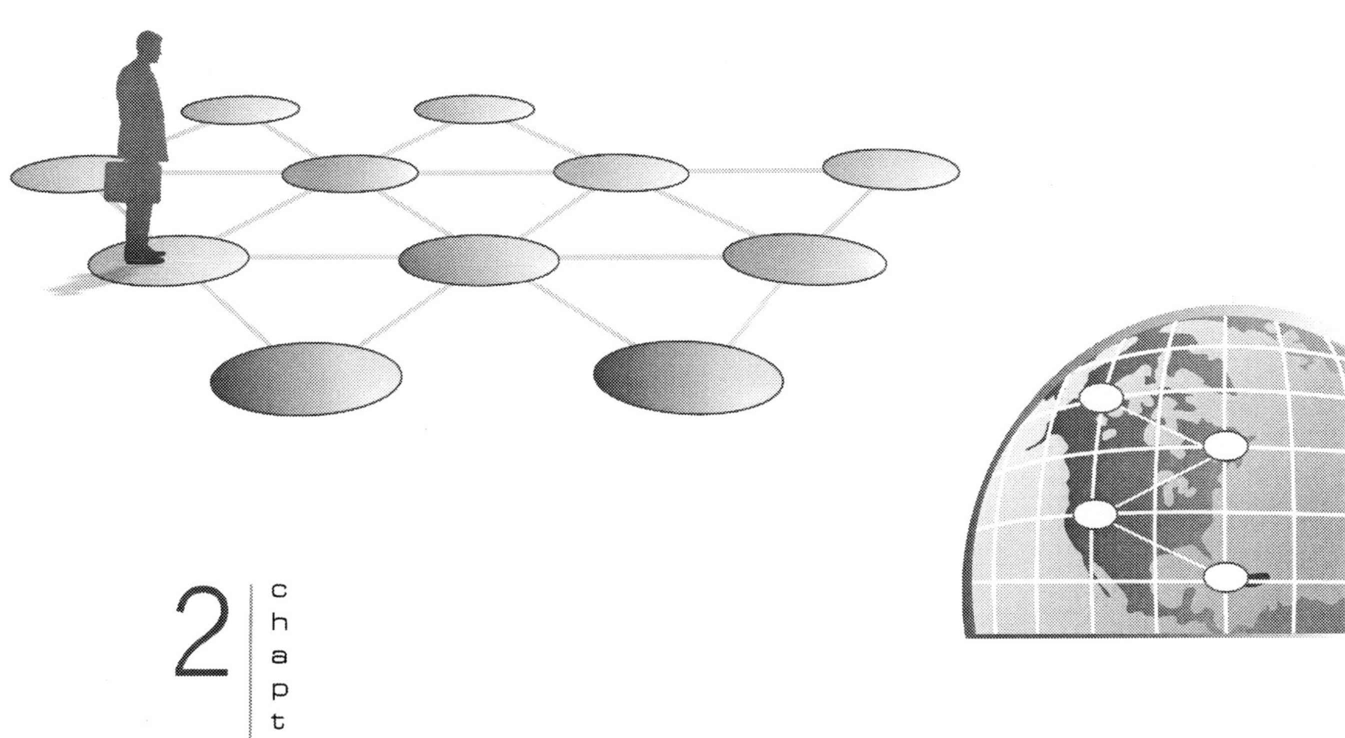

The importance of brands

Branding is not just a case of placing a symbol or name onto products to identify the manufacturer. A brand is a set of attributes, that have a meaning, an image and produce associations when being considered by a consumer.

Branding is probably one of the oldest theories of marketing with evidence throughout the world from ancient Rome, Greece and India as early as 1300 B.C. Within early barter societies, potters would place distinctive markings on pots so that the craftsmen could be identified. Clearly this is the same as placing a logo, the implicit point however was that only pots of the highest quality where the producer had a great reputation would be sought.

Within this chapter we look at why the concept of branding is important. Initially we compare the difference between a brand and a product. We then look at the opportunities for brands to develop and what components of brands exist. Fundamentally, as with our potter example, brands are there to build customer loyalty. Aspects of loyalty are discussed before we look at the concept of brand personality and how this might help position a brand in the mind of consumers to represent something that is of significance, relevance and value to them.

Contents

By the end of this chapter you will be able to:

- Discuss the definition of a brand and how this differs from a product
- Consider the various opportunities for branding in terms of what can be branded
- Identify and deconstruct elements of a brand
- Evaluate different aspects of brand loyalty
- Explain the concept of brand personality
- Explore the meaning of brand positioning
- Identify the concept of brand value

1 What is a brand?

A **brand** is a name, term, sign, symbol or design intended to identify the product or service of a seller and to differentiate it from those of competitors (Aaker, 1991). Not long ago, and this is still the case in many less developed countries, most products were sold unbranded from barrels and bins etc. Today in developed and even developing countries hardly anything goes unbranded. Salt, oranges, nuts and screws are often branded. There has however been a limited return recently in some developed countries to 'generics'. These are cheap, unbranded products, packaged plainly and not heavily advertised.

A product is something that is made in a factory and a service is an activity carried out; a brand is something that is bought by a customer. A product (and often a service) can be copied by a competitor; a brand is unique. Branding is about building a unique identity which can be protected and sustained against competition. Kolind (2009) eloquently describes the difference:

"You often come across two products which when compared on ingredients are identical, when compared on the packaging are identical and when tested in consumer blind tests are considered identical. Still, the consumers are willing to pay considerably more for the version with the known brand. A product can be "copied" but a real brand cannot."

A product can be quickly outdated; a successful brand is timeless (King, 2007).

> **Definition**
>
> A successful **brand** is an identifiable product, service, person or place, augmented in such a way that the buyer or user perceives relevant, unique sustainable added values which match their needs closely (de Chernatony and McDonald, 2003).

1.1 Levels of a brand

The differences between a brand and a product may also be reflected as different levels. The four commonly cited levels are:

- **Generic** – commodity form which meets buyers' or users' basic needs. This is a basic product version

- **Expected** – satisfy target consumers' minimum purchase conditions

- **Augmented** – more refined with added values to meet non-functional emotional needs

- **Potential** – only creativity limits the brand's potential level (what it could be)

(de Chernatony and McDonald, 2003).

Keller (2008) precedes these four levels with a first level titled the 'core benefit level'. At this level there is merely a fundamental need or want that consumers satisfy by consuming the product or service. For example drinking safe tap water may represent a fundamental want.

1.2 The atomic model of the brand

de Chernatony (2006) explains that **brands are clusters of functional and emotional values** and as such are valuable to both organisations and customers. He argues:

"Brands don't just command respect because of their value to corporations. They do so because they add quality to life." p.4

To clarify what is the essence of a brand, de Chernatony (2006) developed the atomic model of the brand. The model distils the components of brands.

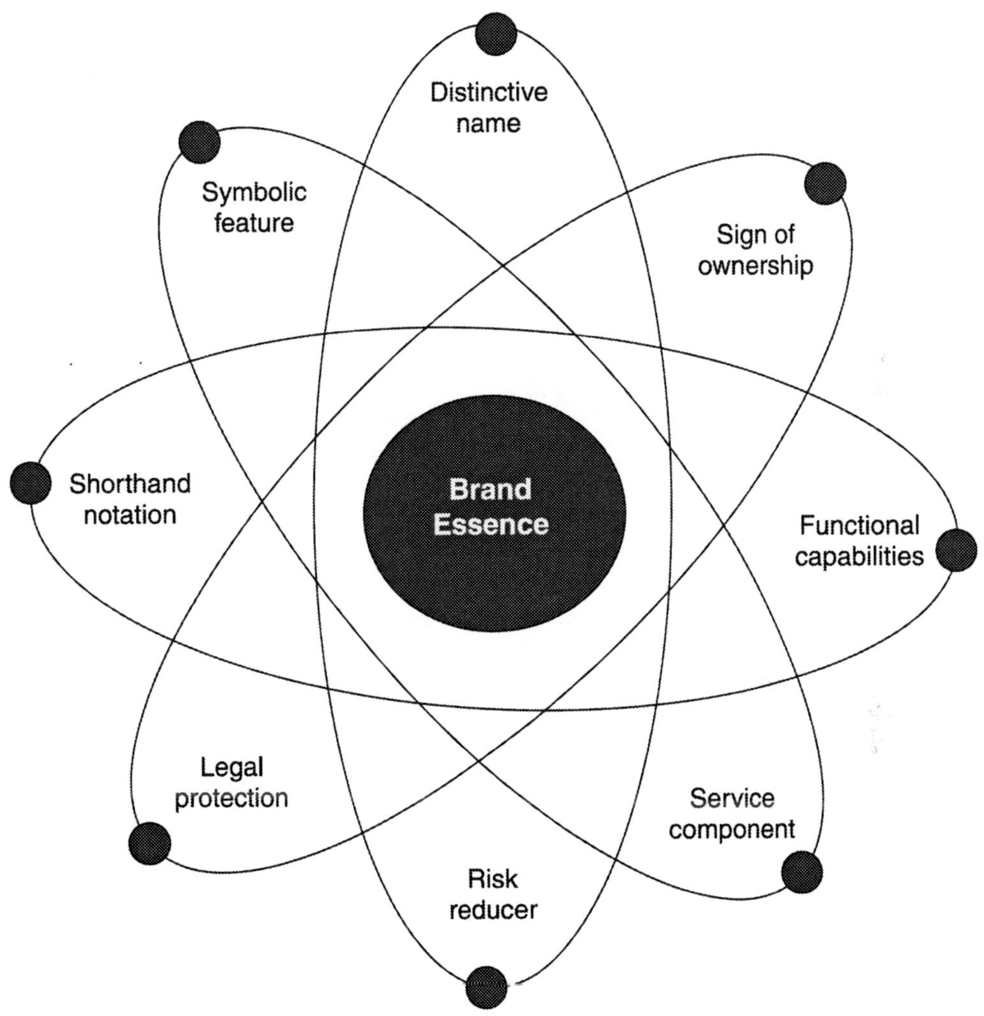

Reproduced with kind permission from Leslie de Chernatony

The model is useful to help identify the elements (electrons) of the brand and to help managers identify where resources should be employed. The elements surround the brand in a trajectory. If the service component for example is the main emphasis within the brand essence, there will be a strong attractive force with the service component electron. The 'electrons' are described in the table below.

Brand essence	Provides clear direction about the nature of the brand. The brand pyramid should characterise the brand essence into a final form which is well resourced.
Distinctive name	Recognisable, easy to pronounce, evokes the benefit of the brand eg Dualit, Huggies, Soothers.
Sign of ownership	Where values of the brand are different from the values of the corporation such as with Proctor & Gamble – Pringles, Aussie, Pampers are often not given a sign of ownership. By contrast companies such as Heinz rely on their umbrella or family brand name. We will return to this when we look at brand architecture in Chapter 3.
Functional capabilities	Factors such as the packaging and communications need to be developed so that the functional capabilities of the brand are recognised by consumers. Ariston washing machines spent decades focussing on the functional qualities of their durability and used the tag line 'Ariston go on and on and on'. Functional qualities are represented by reference to performance, features, reliability differentiated elements and durability.
Service component	Even product brands have services attached to them. These can be in the form of customer care-lines or delivery methods. Service brands frequently compete on their superior levels of customer service.
Risk reducer	Understanding the perceived risks within a purchase scenario enables marketers to position the brand in order to minimise or alleviate concern.
Legal protection	Trademark registration is often emphasised by brands especially those which need to protect their features. Coca-Cola successfully trademarked the shape of their iconic bottle.
Shorthand notion	Through the investment in the brand positioning and personality consumers understand what brands they relate to stand for. This means that they are able to recall the brand qualities or relevance to a given purchase scenario more easily. The brand then becomes a short cut device within the decision making process. To test this for yourself imagine what it is like to shop for groceries overseas. Apart from the language barrier, think how much longer it would take you to shop.
Symbolic feature	The badge value of the brand. Some brands enable consumers to communicate something about themselves. Often this is linked to emotion or status. Consumers personify brands and when they look at the symbolic values of brands they select those which match their actual or desired self concept.

2 Branding opportunities

According to Kevin Lane Keller (2008):

"virtually anything can be and has been branded" p. 27

Although branding is an ancient concept the modern day theory began within the realms of the FMCG sector. Some of the world's best known brands were established in their country of origin often over a hundred years ago.

Kevin Lane Keller (2008) suggests the following as a discussion question. Try working through this question for yourself.

Can you think of yourself as a brand? What do (or would) you do to 'brand' yourself?

You might want to review this activity again once you reach the end of the chapter. Does your view change?

Keller (2008) discussed the range of industries where branding is an integral part of marketing strategy. The table below summarises his findings.

Sector	Brand strategies
Business to business products	Creating positive reputation and image for the company Goodwill generation leads to greater selling opportunities and relationships
Hi-tech products	Have struggled with branding in the past Success is no longer driven by innovation Speed and brevity of technology product life cycles cause challenges for branding. Trust is critical.
Services	Last 10 years have seen services brands accelerate Challenge due to their intangible nature Branding helps to address lack of tangibility and variability issues
Retailers and distributors	Brands generate consumer interest, patronage and loyalty in stores. Own brands can be introduced
Online products and services	Initially web marketers oversimplified the online branding process due to overly flashy sites which failed to lead to awareness about what the product or service actually was. Real time customer service agents are not increasing which has led to an improved perception of online customer service Important to use off-line activities to draw traffic Benefits of interactivity, customisation and timeliness only just being realised.
People and organisations	As they are 'real' people branding is relatively straightforward because they have well defined images. By building up a name and reputation in a business context – you are building a brand. Some organisations take on meanings through programmes and activities. This is especially true of non-profit organisations such as UNICEF, Red Cross, Amnesty International.

Sector	Brand strategies
Sports, art and entertainment	Sports marketing has become increasingly sophisticated using traditional FMCG branding techniques. Symbols and logos are especially important as are licensing agreements and the related financial support.
Geographical locations	Tourism marketing has led to a number of cities, regions, states and countries actively promoted with a variety of communication tools.
Ideas and causes	Slogans, coloured ribbons, badges, car stickers, and even songs are used in order to brand ideas especially those supported by non-profit organisations. Politicians attempt to use these methods during political elections.

3 Brand elements

The essence of a brand can be understood through the use of two well know models. Brand Pyramids and the Prism of Brand Identity.

3.1 Brand pyramids

There are many conceptions of brand pyramids, Keller (2008) discusses brand building blocks combining to create a pyramid. These blocks represent both rational and emotional associations with brands and managers can choose while building their brands to reflect either one or both aspects. The building blocks are shown in the diagram below. The table outlines the stages of brand development and the objectives set for the brands at each stage.

Pyramid block	Focus	Stages of brand development	Branding objective at each stage
Salience	Category identification Need satisfied	Identity – who are you?	Deep, broad brand awareness
Performance	Primary characteristics and secondary product reliability features Durability and serviceability Service effectiveness Efficiency and empathy Style and design Price	Meaning – what are you?	Points of parity and difference
Imagery	User profiles Purchase and usage Situations Personality and values History and heritage Experience		

Pyramid block	Focus	Stages of brand development	Branding objective at each stage
Judgements	Quality		
	Credibility		
	Consideration		
	Superiority		
Feelings	Warmth	Response – what about you?	Positive accessible reactions
	Fun		
	Excitement		
	Security		
	Social appeal		
	Self respect		
Resonance	Loyalty	Relationships – what about you and me?	Intense, active loyalty
	Attachment		
	Community		
	Engagement		

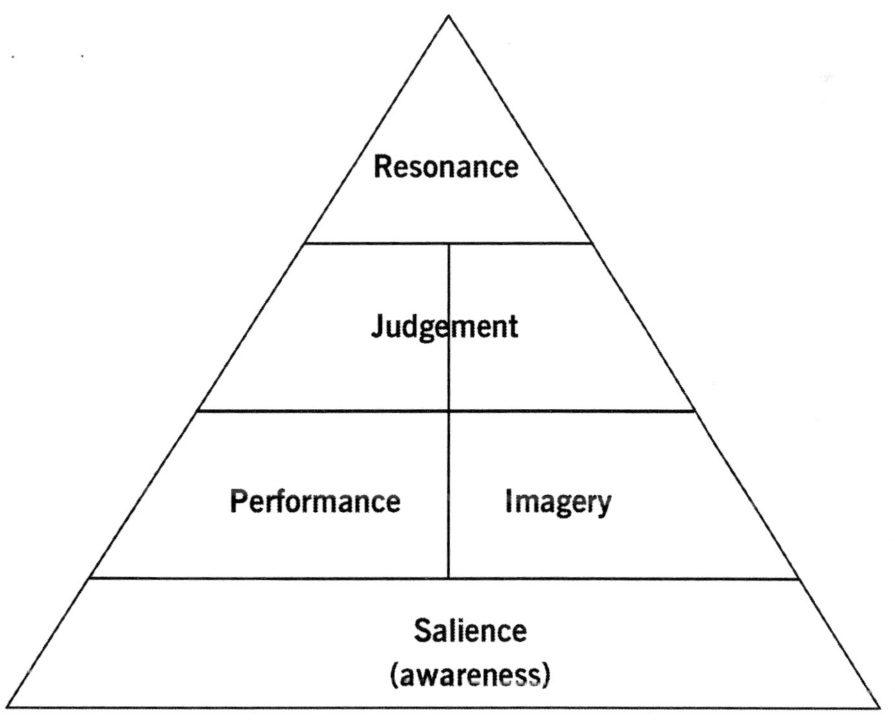

Source: Adapted from Kapferer (2004).

We will return to brand pyramids at the end of the chapter when we look at the concept of brand equity.

3.2 The Prism of Brand Identity

Kapferer's (2008) Prism of Brand Identity outlines the totality of brand essence. Kapferer (2008) states that this represents the core of the brand. There are elements of identity which must remain permanent and parts which can evolve with brand strategy. Brands have the gift of speech and can only exist (as metaphorical personalities) if they communicate.

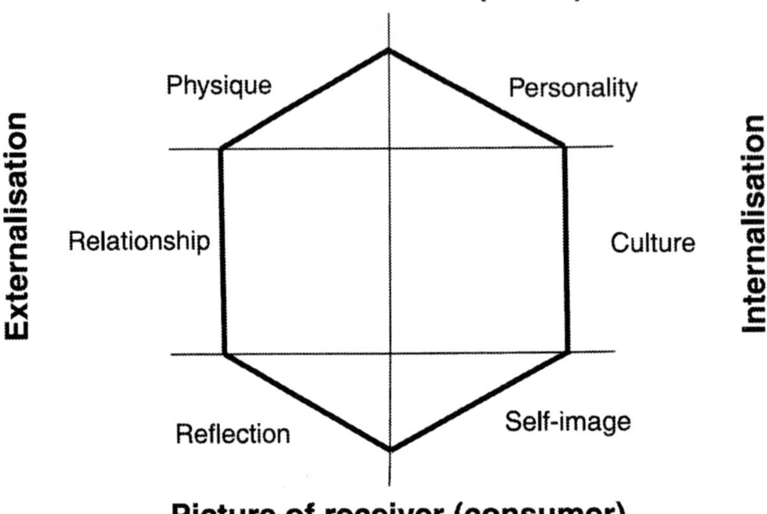

Adapted from Kapferer (2008)

The brand identity prism represents the brand through the six sides of the prism.

Brand identity prism	Meaning
Physique	The basis of the brand. The physical aspect eg does the brand represent a product or a service?
Personality	The character of the brand. Aaker (1997) refers to personality as the set of human characteristics associated with a brand. We will return to personality later in the chapter.
Culture	Country of origin, organisational culture and the brand values. The culture provides the inspiration to manage brands.
Relationship	The relationship between the consumer and the organisation as an intangible exchange.
Reflection	Consumers' perception of what the brand stands for and how this reflects them. It refers to how the consumer wishes to be seen rather than as they actually are.
Self-image	The consumers' view of self and how these lead to attitudes about the brand. Brands are viewed according to how they fit or help build the consumers' self-image to the outside world. It is the brand element which explains conspicuous consumption.

External aspects of the brand refer to the aspects that are viewed by those outside of the organisation and act as the brand 'face'. These include:

- Name
- Logo
- Advertising
- Brand and corporate identity
- Products and service levels

Internal aspects of the brand refer to processes and systems in operation within the organisation which help to give the brand life. These include:

- Brand values
- Quality
- Staff motivation

- Skill and training of staff
- Customer relations
- Business processes

Kapferer (2004) argued that there is more than one perspective from which to study the brand. The model looks specifically at the brand from the viewpoint that it communicates, and that consumers in turn communicate with the brand. It is important to consider the view of the sender (the brand in terms of its physique and personality and the receiver who is the consumer (in terms of reflection and self-image). The relationship and cultural aspects bring the consumer and the brand together and are relevant to both parties.

Activity 2

Work through the Prism of Brand Identity and try to establish the elements for a brand you are familiar with.

3.3 Application of the models

The concepts of the brand pyramid and identity prism are used for a number of purposes:

- To enable management and their agencies to understand the brand, its strengths and opportunities.

- To develop brand strategy and the formulation of the brand's positioning in the market.

- To enable the brand team to develop consistency in the message being transmitted through packaging and design, advertising, below-the-line activities and through potential brand extensions.

- To understand the brand's core and style helps to determine how far the brand can be meaningfully stretched to other products and market segments.

(Doyle, 1998 pp.173–174)

The Brand channel has a link to a useful literature review on brand building. It can be accessed at:

http://www.brandchannel.com/images/papers/257_A_Brand_Building_Literature_Review.pdf

4 Brand loyalty

The concept of brand loyalty is also a well entrenched marketing theory. Yet, both academics and practitioners recognise its importance but also that it is a complicated construct. There is much disagreement about how to measure loyalty. Alternative measures suggested include:

- Satisfaction
- Liking
- Commitment
- Switching costs
- Promises of repurchase

Researchers are split about which is the best measure.

4.1 Types of loyalty

Loyalty does differ according to:

- attitude strength of consumers
- the extent consumers perceive brand differentiation
- whether it is true or false loyalty
- customer disposition – in terms of whether they are habitual, loyal, variety seekers or switchers
- alternative segments and customer groups

Differences in loyalty have often led to the assumption that people are disloyal defectors if they try competitors' brands. In reality, consumers have portfolios of brands which they rotate according to their differing needs and purchase scenarios (Knox, 1997). For example if I am entertaining friends I may choose a premium brand of wine rather than a cheaper retailer own or non-branded variant. This will be for a number of reasons, one being the projection of my own self concept and two that I would want my friends to enjoy the wine so a recognised brand will remove part of the risk of not knowing the quality of the product. If the wine were just for myself, I would be more likely to take a risk with a cheaper alternative as there would be no social risk associated with the purchase. The concept of conspicuous consumption is therefore closely related to brand purchase motivations.

Activity 3

Think about the brands you use regularly, try to think about:

- How you would describe their personalities
- The sort of consumer-brand relationship you may have formed with the brand.

We will discuss the concept of consumer-brand relationships in section 5. Brand relationships can answer questions about brand portfolio usage and loyalty. Regarding the metaphorical concept of consumers entering relationships with brands enables us to recognise that there are many different relationship *types* that consumers can have with brands. In everyday life people have an assortment of relationships within different contexts and situations and the same can be said of the brands that they form partnerships with. Gifford (1997) pointed out that to say a consumer is brand loyal is like saying you will marry everyone you meet or else they will never be a meaningful part of your life.

It is through understanding the concept of brand personality that we are able to acknowledge brand level relationships (Fournier, 1996).

5 Brand personality

The concept of brand personality is well established within the field of brand studies. Although it is essentially a metaphor (in most cases, although we have already made the case for people becoming brands in their own right), it is powerful and encourages the development of emotional bonds and meaning (Aaker and Fournier, 1995).

The techniques used to instil a brand with personality are:

- **Anthromorphisation** – attributing human qualities to an animal or inanimate object

- **Personification** – creating a tangible 'puppet' eg brand mascots such as The Milky Bar Kid, Bibbendum, Ronald McDonald

- **Imagery** – using emotional imagery

These methods essentially put the marketer in the position of a puppeteer (Vincent and de Chernatony, 1999) with the brand manager responsible for the strings. Brand personality encourages the

development of emotional bonds and meaning. Consumers tend to like brands which reflect their own personality whether this is their true or aspirational persona. For example if somebody regarded themselves or wanted to be regarded as wealthy and influential they may be more likely to choose a Rolex watch over a Swatch.

Think for a moment about the number of successful teddy bear and doll brands. Not all of these are designed exclusively for children and in some countries (Japan being one) young adult women are key targets for some toys such as Miffy, Strawberry Shortcake and others. Ty Beanie babies was a huge success story of the 1990s, in a similar way to Cabbage Patch Kids in the 1980s. The reason for their success was partly due to the naming and birth certificate 'add ons' which helped to provide a story and history to surround the toy's personality. The ability to instill distinct personalities enables a stronger relationship between the brand and the consumer.

The Internet has certainly assisted with helping to strengthen brand personality. Consumers are more able than ever to find out increasing amounts of information (both good and bad) about the brands they use.

Brand personality is the key to brands and consumers actually building relationships. Relationships built between consumers and their brands are inherently a result of the development of strong brand personalities. If brands have a personality then consumers may not just perceive them but also have relationships with them. As this relationship develops, stronger bonds arise and repeat usage may occur. This is where branding becomes a key aspect of relationship marketing strategies.

For a brand to become a relationship partner it must surpass personification and behave as an active partner.

5.1 Consumer-brand relationships

Just as we form different forms of relationships in our interpersonal lives, we form different relationships with our brands. Fournier (1994) identified **metaphorical relationship types** consumers may develop with brands. These included **friendship** (casual, childhood friends), **romantic** (courtship, committed partnerships, flings) and **dark side relationships** (dependency, enslavement). Individuals do not enter relationships with every other person that they meet and certainly not the same relationship with everyone. King (1984) explained how consumers choose brands as they would friends:

"there are degrees of friendship, and people rarely stick exclusively to just one friend; in the same way there are degrees of brand loyalty, very rarely loyalty to one brand alone. Friendships come and go, according sometimes to what friends say and do. Not everyone will like the same person, some are more popular than others" p. 12

Casual friends (as opposed to committed partnerships and childhood friendships) tend to also change according to one's life stage. This too is the same with brands. Consumer life stage models focus on this area. There are some brands that we encounter as children that we remain committed to throughout our lives. Others we purchase according to our needs, influential third parties and purchase capabilities at the time.

When positioning brands, marketers look to create the appropriate brand personality and associated cues as required to appeal to the target market and thus attempt to build positive relationships.

6 Brand positioning

> **Definition**
>
> **Positioning** is defined as 'the place the product [or organisation] occupies in consumer's minds relative to competing products [or organisations]' or 'the way the product [or organisation] is defined by consumers on important attributes' (Kotler & Armstrong, 2000).

An organisation will seek to determine how its corporate image (and/or brand, if relevant) is perceived by stakeholder audiences in relation to its competitors. For example, consider Diageo's positioning statement: 'Diageo is the world's leading premium drinks business' (www.diageo.com). This has the virtue of clarity, consistency (over several years' use), credibility (because it is backed by brand performance) and competitiveness (positioning the brand as superior to its rivals).

You should be familiar with the use of perceptual maps or positioning charts to plot competing brands against two or more dimensions (perceived attributes). The same techniques can be used to position company reputations or images, or corporate brands:

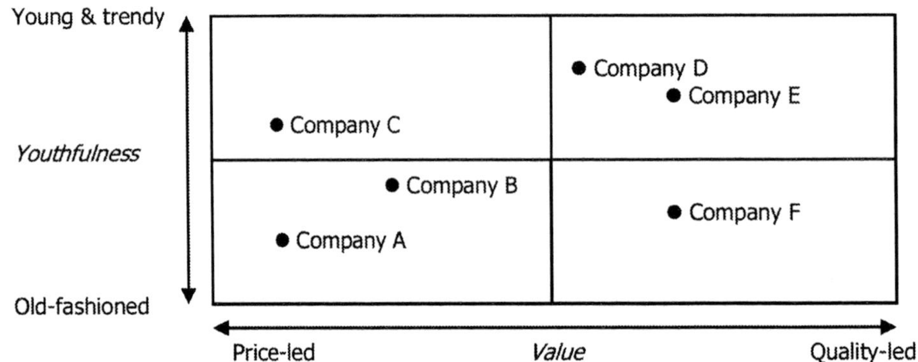

Corporate brand/image positioning map

In this generalised example, Company A may have a valuable reputational niche among older, price-conscious consumers. On the other hand, it may note that there is a gap in the market to engage older, or more conservative but more affluent, quality-seeking consumers.

The strength of a brand's positioning in the marketplace is based on elements (Jobber, 2007) such as:

- **Brand heritage**: the background 'story' to the brand and its culture (eg the long retail history of Marks & Spencer).

- **Brand values**: the core values and characteristics of the brand (eg The Body Shop's ethical values, Audi's sophistication and progress, Virgin's entrepreneurship and fun). We return to these in section 7.

- **Brand assets**: what makes the brand distinctive from other competing brands: symbols (eg the Nike 'swoosh'), attributes (eg 3M or Dyson UK's innovation) and relationships (eg Dell Computers' late customisation and direct selling network).

- **Brand personality**: the 'character' of the brand, often described in terms of other entities, such as people, animals or objects. (Eg 'If Organisation X were a person, what kind of person would it be?')

- **Brand reflection and identification**: how the brand relates to stakeholder self-identity, or how stakeholders perceive themselves as a result of engaging with the brand (eg as an Apple user, a John Lewis employee or a Shell investor).

- **Brand awareness**: developing credibility and raising visibility are important tasks in brand management. Brand awareness objectives will depend on the industry and markets, and the organisation's position within them (Aaker, 1996).

    - **Top of mind** awareness (or brand salience) requires establishing the brand as one people know without reference to an industry or a prompt (eg Virgin, BBC, Greenpeace)

    - **Brand recall** awareness requires a context prompt (eg industry or category)

    - **Brand recognition** awareness requires the brand being recognised when mentioned.

As environments and stakeholders change and/or the organisation changes, the brand may need to be repositioned. Nokia, for example, re positioned itself from being a paper manufacturer to a market leader in mobile phones.

Brand consultancy Interbrand offers a range of interesting brand-related articles, free to download, on subjects such as the impact of sustainability on corporate brand value, and the kinds of corporate messaging required in an economic downturn.
Link: http://www.interbrand.com   (Follow the link > Knowledge > Papers and articles.)

6.1 Re-branding

There may be occasions when an organisation needs to reposition itself in its market, and a radical change of identity is required to throw off old associations and perceptions in stakeholders' minds. This may involve a change of identity and communications (eg McDonalds highlighting sweeping changes with its global 'I'm Lovin' It' campaign).

Global case study

Global brand **McDonalds** was losing ground to competitors by 2001, due to the perception of poor service, boring food – and bad publicity arising from the obesity 'epidemic' associated with fast food. It launched a 'Plan to Win' **total rebranding campaign**, designed to woo back core customers: mothers, kids and 20-somethings. McDonalds recognised that the brand was losing relevance, because it had focused on opening new outlets – rather than maximising sales at existing outlets. The worldwide 'I'm Lovin' It' campaign, featuring celebrities like Justin Timberlake, Destiny's Child and Yao Ming, has **repositioned** McDonalds as an energetic lifestyle brand. Service and outlet décor have been revamped. New health-conscious menus have been introduced – at a profit-raising premium. The results? Same-store results up 7% worldwide. Complaints down 11%. Compliments up 18%. (*Business*, Jan/Feb 2005)

However, since the brand name is the encapsulation of the brand image – and all the associations that go with it – rebranding is often identified with **change of name**: if the imagery is changed but the brand name remains the same, customers may become confused and alienated.

Change of name may also be helpful or necessary:

- In overcoming identity crisis following **merger or restructuring**, especially if there are internal political/power issues over 'ownership' of the merged or newly autonomous brand. When a merger or acquisition takes place, for example, the corporate name may reflect the dual identity as when Glaxo Wellcome and SmithKline Beecham merged to become GlaxoSmithKline (GSK). Or it may be a completely new name, to avoid the difficulties of aggregation (eg when drinks companies GrandMet and Guinness became Diageo). Another interesting example is Andersen Consulting's change of name to Accenture, as a result of a rift with parent accounting firm (Arthur Andersen) in the 1990s. (An interesting exercise in reputational risk management, since Arthur Andersen was dissolved in 2002 following its involvement in the Enron scandal...)

- In disassociating the corporate brand from **negative or old-fashioned images**. BT Wireless, for example, changed its name to O$_2$: away from an old-fashioned, bureaucratic image, and towards a youthful image (based on focus group statements that they saw their mobile phones as essential to life, like oxygen). Similarly, Kentucky Fried Chicken rebranded to KFC because of the negative connotations of fried food.

- In disassociating the corporate brand from negative perceptions of an **industry or country of origin** (where used in the name) or associations with the name itself. An example would be the Philip Morris tobacco companies adopting the name Altria Group to disassociate their (at the time) food business (Kraft) from the negative perceptions of tobacco companies.

- In reflecting corporate **diversification** out of its original product or business category, making the original name limiting or inaccurate. For example, Esso (Standard Oil) changed its name to Exxon, as its portfolio extended beyond oil.

- In taking advantage of **subsidiary brand strength**. For example, Consolidated Foods discarded its unfamiliar name and adopted the name of its leading brand: Sara Lee.

- In creating **global brands**, by harmonising brand names across international markets. For example, the UK One2One brand changed its name to T-mobile (as used by its parent company Deutsche Telecom in Germany). We cover this area in more detail in Chapter 4.

Activity 4

Who will be the primary audiences for a re-branding communication programme, and how might they be reached?

---

van Riel and Fombrun (2007) offer the following general points about re-branding programmes.

- Corporate branding initiatives usually develop as a result **stakeholder pressure for increased clarity** and transparency of corporate communication – which may be resisted by managers.

- **Strong and assertive leadership** is required to develop and implement corporate branding.

- The (re)launch of a corporate brand must be supported with **strong symbolic support**, internal and external communication and 'hoopla'!

- (Re)branding **generates resistance**. Care must be taken to develop and agree rules for business units to follow (eg in regard to visual identity), and to monitor their implementation.

Rebranding presents considerable risks and challenges, where stakeholders have built up powerful engagement with an old brand.

Global case study

In the UK, The Post Office's attempt to re-brand as Consignia, for example, was short-lived, due to objections from consumers, employees and the media. The decision was reversed and the corporate name 'Royal Mail Group' selected, forming a closer association with the former brand.

For more detail on the branding and re-branding process, if this is relevant to your organisation and assessment tasks, you may wish to browse through some of the supplementary readings in this area. See:

Kapferer, J. (2008) *The New Strategic Brand Management*. 4th ed. London: Kogan Page. This is good all round on the subject, but see particularly Chapter 5: Brand diversity: the types of brands, which explores countries, towns, universities, celebrities and other non-corporate entities as brands; and Chapter 13: Brand architecture.

There are many examples of brand names changing in order to provide global unification (Snickers) or following a merger (Diageo). We will look at the global brand issues later in Chapter 4.

Research agency Millward Brown have written a useful paper on how to change a brand name successfully. It can be downloaded from the following address: http://www.millwardbrown.com/Sites/MillwardBrown/Media/Pdfs/en/KnowledgePoints/306AC40F.pdf

7 Brand values

Creating a successful brand can be very expensive in money and effort. When a company makes a significant investment in creating and enhancing a brand, it is appropriate to measure the degree of success that it achieves. This is the process of **brand valuation**.

You must be careful to distinguish between brand value and brand values. Brand values are the intangible factors that make up the identity or image of the brand. They are usually quoted as a list of adjectives, such as young, active, exclusive, fun, reliable, secure, safe and so on. Brand value (valuation), on the other hand, is value in an economic sense: the monetary worth of the brand, or the cash value it represents. We will look at brand valuation in Chapter 4.

Branding needs to add value so that the consumers perceive a meaning in a brand that is relevant to them (this is the key argument for resourcing brands). Value can be achieved through the way buyers perceive the performance of the brand, the psychosocial meanings attached to a brand and the level of brand name awareness.

Adding value is a core concept as in order for consumers to engage with brands they have to believe that the brand offers something of significant value. We looked earlier at the concept of brand personality and noted that consumers relate to brands which represent their actual or ideal self concept.

As we have suggested already within this chapter, a number of meanings are then attributed to brands. These can be emotional and psychological meanings. Ferrari for example will evoke a perception of wealth which in turn may trigger other emotional and psychological responses either consciously or subconsciously. For different individuals this will result in a slightly different effect, for some it may even be a negative connotation depending on their own personal, cultural, emotional and lifestyle situations.

GMN viewpoint

Crispin Reed is the Managing Director of Brandhouse and Advisory Council member of GMN.
Brandhouse specialise in identifying the emotional attachments consumers make with their brands.

Emotion 100 is a survey of 100 brands where respondents are asked to rate top brands according to seven emotional principles. These are:

- Belonging
- Commitment
- Enjoyment
- Pride
- Compassion
- Desire
- Excitement

The 2009 survey found that the even during times of recession where it was anticipated that consumers would ditch brands in favour of cheaper retailer own label varients, those brands which had a high emotional attachment were the least likely to be culled from a shopping basket.

Difererent products and services are likely to evoke different patterns of resonance with their consumers. For instance, not suprisingly the car brands had significantly higher resonance with men than women.

The top five ranking brands in terms of their emotional connections were:

- Google
- Sony
- Heinz
- BBC
- Kellogg

The report as published in Marketing can be accessed from:

http://www.brandhouse.co.uk/brandhouse/sites/default/files/pdf/Emotion%20Sells.pdf

7.1 Brand equity

What exactly is brand equity?

Definition

Brand equity, as first defined by Farquhar (1989), is *"the 'added value' with which a given brand endows a product"* p.24.

Apart from Farquhar's first definition of brand equity, other definitions have appeared. According to Lassar, Mittal and Sharma (1995), brand equity has been examined from a financial (Farquhar, 1989; Kapferer, 1997; Doyle, 2001), and a customer-based perspective (Keller, 1993).

In other words, financial meaning from the perspective of the value of the brand to the firm, and customer-based meaning of the value of the brand for the customer which comes from a marketing decision-making context.

Brand equity has also been defined as *"the enhancement in the perceived utility and desirability a brand name confers on a product"* (Lassar, Mittal and Sharma, 1995 p.13).

High brand equity is considered to be a competitive advantage since it implies that firms can charge a premium; there is an increase in customer demand; extending a brand becomes easier; communication campaigns are more effective; there is better trade leverage; margins can be greater; and the company becomes less vulnerable to competition (Bendixen, Bukasa, and Abratt, 2003). In other words, high

brand equity generates a "differential effect", higher "brand knowledge", and a larger "consumer response" (Keller, 2008), which normally leads to better brand performance, both from a financial and a customer perspective.

The added value conferred by a brand is largely subjective. In blind testing many consumers cannot tell the difference between different products (Coca-Cola and Pepsi are a classic example); however, they will exhibit a preference for a strong brand name when shown it. Apart from the quality and functionality of the product, brand equity is built on suggestion rather than substance.

A useful article on brand equity and how this relates to brand preference and purchase intent as published in the Journal of Advertising can be found at:
http://www.allbusiness.com/marketing/market-research/528470-1.html

Brand equity is the asset the marketer builds to ensure continuity of satisfaction for the customer and profit for the supplier.

Most consumer buying decisions, therefore, **do not depend on the functionality of the product**.

(a)   Products are bought for **emotional reasons**. For example, most sports trainers are fashion products.

(b)   Branding reduces the need for the intellectually challenging process of rational choice.

7.1.1 Sources of brand equity

Experience	Customer's actual usage of a brand can give positive or negative associations.
User associations	Brands get an image from the type of people using them; brands might be associated with particular personalities.
Appearance	Design appeals to people's aesthetic sensibilities.
Manufacturer's name	The company reputation may support the brand.
Marketing communication	Building the brand by establishing its values is a major reason why marketing communication of all kinds is undertaken.

1 Discuss the definition of a brand and how this differs from a product

- Brands are unique.
- A product with unique sustainable added values which match needs closely
- Levels – generic, expected, augmented and potential.
- Atomic model of branding has the following elements: distinctive name, sign of ownership, functional capabilities, service components, risk reducer, legal protection, shorthand notion and symbolic features.

2 Consider the various opportunities for branding in terms of what can be branded

- Virtually anything can be and has been branded.
- Relevant to all industries.

3 Identify and deconstruct elements of a brand

- Brand pyramids are used to represent rational and emotional associations with brands.
- Brands can reflect either rational or emotional elements or both.
- Brands can only exist when they communicate

4 Evaluate different aspects of brand loyalty

- Disagreement about measures of loyalty exist.
- Consumers tend to have portfolios of brands used according to different needs and purchase scenarios

5 Explain the concept of brand personality

- Projecting a human personality on a brand helps consumers to identify with the brand.
- Consumers purchase brands which match their own or ideal personality
- Consumer-brand relationships can form. These also explain differences in loyalty in given situational contexts.

6 Explore the meaning of brand positioning

- The place the brand occupies in the consumer's mind relative to competing products.
- Developed as a result of the combination of brand elements
- Rebranding is sometimes necessary.

7 Identify the concept of brand value

- Brand values are the intangible factors that make up the identity or image of the brand

1   Individuals can be brands. Think about the notion of celebrity brands, they are simply 'people brands' which may have worked a little more at their positioning and targeted communications with their audiences. You also will have audiences and as such will mean something to those individuals. Later in the chapter we cover brand elements and positioning, so work through your answer with these aspects in mind.

2   This will depend on the brand you have chosen. If we look at the example of sports brand Adidas however we can distil the following key aspects of the brand which might be represented by the prism.

Brand identity prism	Adidas example
Physique	Sport and fitness
Personality	Traditional, conservative, collective
Culture	European, traditional
Relationship	Heritage and quality
Reflection	True sportsmanship and team playing
Self image	Relates more to competing than winning

3   This will depend on your own research.

4   This will depend on your own research.

Aaker, D.A., (1996). *Building Strong Brands.* New York: The Free Press.

Aaker, J. and Fournier, S., (1995). A Brand as a Character, a Partner and a Person: Three Perspectives on the Question of Brand Personality. *Advances in Consumer Research*, vol 22 pp.391-395.

de Chernatony, L., (2006). *From Brand Vision to Brand Evaluation.* 2nd edition. Oxford: Elsevier Butterworth-Heinemann.

de Chernatony, L. and McDonald, M., (2003). *Creating Powerful Brands*. Oxford: Butterworth-Heinnemann.

Doyle, P., (1998). *Marketing Management and Strategy*. London: Prentice Hall.

Farquhar, P., (1989). *Managing Brand Equity*. Marketing Research vol 1 issue 3 p. 24-33.

Fournier, S., (1996). *The Consumer and the Brand: An Understanding within the Framework of Personal Relationships.* Working paper. Boston: Harvard University.

Gifford, D., (1997). *Brand Management: Moving Beyond Loyalty.* Harvard Business Review. March-April, pp. 9-10.

Kapferer, J., (2008). *The New Strategic Brand Management.* London: Kogan Page.

Keller, K., (2008). *Strategic Brand Management.* International edition. 3rd edition. Harlow: Prentice Hall.

King, S., (1984). *Developing New Brands.* London: J Walter Thompson.

King, S., (2007). *A Master Class in Brand Planning.* London: Wiley.

Kolind, J., (2009). *The Mixed Brand and Private Label Strategy: Illustrated by the Rema1000 Case.* Conference proceedings of the 25th IMP conference, Marseille, France.

Kotler, P. & Armstrong, G., (2000). *Principles of Marketing.* 9th edition. Englewood Cliffs, N J: Prentice Hall

Kotler, P. Armstrong G., Meggs D., Bradbury, E. & Grech, J., (1999). *Marketing: An Introduction.* Melbourne: Prentice Hall Australia.

Tuominen, P., (1999). *Managing Brand Equity.* LTA, (1), 99, pp. 65-100

van Riel, C. and & Fombrun, C. J., (2007). *Essentials of Corporate Communication: Implementing Practices for Effective Reputation Management.* Abingdon, Oxon: Routledge.

Vincent, K. and de Chernatony, L., (1999) *Investigating Relationships at the Level of the Brand from the Perspectives of Both Organisations and Consumers.* Paper presented at 15th IMP Conference, Dublin.

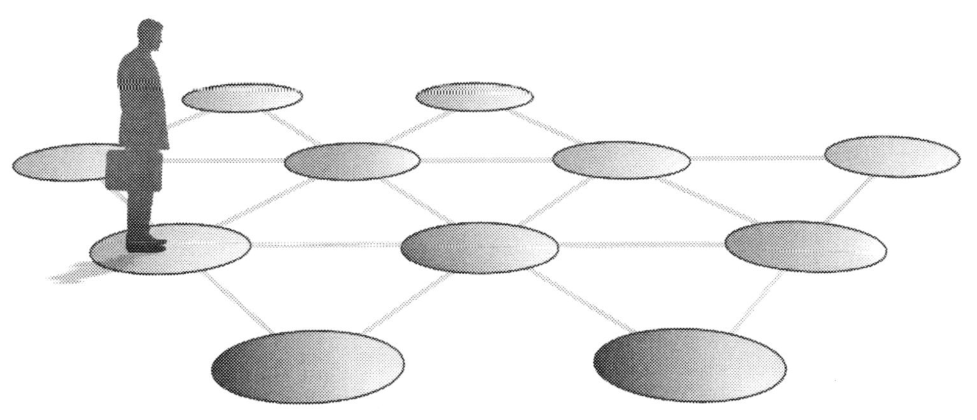

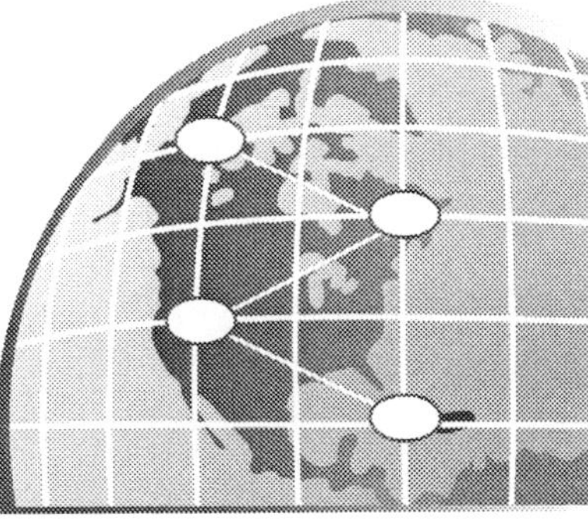

3 chapter

Designing and implementing brand strategy

How to build brands so that they thrive and prosper is an area where many consultancies excel at selling proprietary techniques.

Essentially strategy design should be consistent and follow a logical planning process. Many of the concepts we developed in Chapter 2 will be considered and it will be worth keeping this chapter in mind as you work through Chapter 3.

While covering this area, bear in mind that brand strategy must successfully integrate with the wider communications and marketing plans. As such tactics employed must reflect this wider remit. The chapter begins by reviewing the benefits of using resources to build brands. What makes a strong brand is then considered.

Brand programmes are used to help develop brand strategy in order to build brand equity. Channel strategies, communication strategies, pricing strategies and other marketing activities can all be used to increase the value of brands to consumers.

The chapter draws to a close by looking at brand architecture – an alternative definition of brand strategy.

Contents

In this chapter you will cover the following:

- Identify the steps to build strong brands
- Coordinate brand programmes
- Design brand architecture

1 Building strong brands

Before considering how to build strong brands it is useful to review the benefits of branding in the first place.

1.1 Benefits of branding

Beneficiary	Benefit of branding
Customers	• Branding makes it easier to choose between competing products, if brands offer different benefits. Brands help consumers cope with information overload
	• Brands can support aspirations and self image
	• Branding can confer membership of reference groups
Marketers	• Enables extra value to be added to the product
	• Creates an impression in the consumer's mind; encourages re-purchase
	• Differentiates the product, especially if competing products are similar
	• Reduces the importance of price
	• Encourages a pull strategy
	• Other products/services can exploit the brand image (eg Virgin)
Shareholders	• A brand is an intangible asset; even though it is not on the financial statements, a strong brand promises to generate future cash inflows and profits. This is called **brand equity**.

	• Brands build market share, which can generate high profits through:
	  – Higher volume
	  – Higher value (higher prices)
	  – Higher control over distributors

Evolution of brands. Brands have evolved over time, to the extent that they satisfy customer needs.

(a)   **Classic brands** (post World War II) were linked to a single goal (eg cleaner clothes).

(b)   **Contemporary brands** meet functional needs but give associated benefits (eg Volvo and safety).

(c)   **Post-modern brands**? Consumers use brands to attain a broad array of goals, as a result of 'time famine'.

> (i)   Some marketers suggest that brands have an emotional content. Certainly, this might be the case for fashion items (eg trainers) where they confer status.
>
> (ii)   Strangest of all is Mercedes (a subsidiary of Daimler Chrysler), previously known for luxury cars. Mercedes now has product offering that embraces small cars.

Keller (2008) has designed a **four step process for building strong brands**. The steps are phrased as questions:

1   Who are you? (brand identity)

2   What are you? (brand meaning)

3   What about you? What do I think or feel about you? (brand responses)

4   What about you and me? What kind of association and how much of a connection would I like to have with you? (brand relationships)

The order of the steps reflects the brand pyramid that we introduced in Chapter 2. The pyramid, if you remember, starts from identity and goes through to meaning.

To add to notion of brand feelings, Keller (2008) outlines ten commandments of emotional branding.

1   **From consumers to people** – Consumers buy. People live.
2   **From products to experience** – Products fulfil needs. Experiences fulfil desires.
3   **From honest to trust** – Honesty is expected. Trust is engaging and intimate. It needs to be earned.
4   **From quality to preference** – Quality is a given. Preference creates the sale.
5   **From notoriety to aspiration** – Being known does nor mean that you are loved.
6   **From identity to personality** – Identity is recognition. Personality is about character and charisma.
7   **From function to feel** – Function is about practical qualities. Sensorial design is about experiences.
8   **From ubiquity to presence** – Ubiquity is seen. Presence is felt.
9   **From communication to dialogue** – Communication is selling. Dialogue is sharing.
10   **From service to relationship** – Service is selling. Relationship is acknowledgement.

Implementing these commandments into the brand strategy requires key decisions to be made about the brand programme and how the brand architecture will be designed.

Activity 1

Work through Keller's 10 commandments and try to identify where your organisation's brands are performing well (and possibly badly).

Strong brands are therefore those where marketers have a good appreciation of the asset types constituting their brands and according to Keller (2008) the relationships developed as a result with consumers. By recognising which aspects of their brands are particularly valued by consumers, marketers have invested and protected these attributes, sustaining their value and maintaining customer loyalty (or positive relationships) (de Chernatony and McDonald, 2003).

According to deChernatony and Mcdonald (2003) brands whose core values are not tampered with are appreciated by consumers who remain loyal because their positioning has been consistently maintained.

Harry's Bar is probably one of the rare brands which has maintained this consistency since it opened in Venice in 1931. Ever since the first opening day, the bar has attracted the international and refined clientele that habitually visit Venice. It is popular with celebrities and movie moguls during the annual Venice Film Festival and as such enhances it's credibility and reputation through the associated celebrity endorsement.

Giuseppe Cipriani opened the doors to Harry's Bar. In a discreet stone building perched along a canal just off Piazza San Marco in Venice, he created a timeless and impeccably appointed establishment. His concept was to serve others as you would want to be served yourself. This vision has remained consistent today and is maintained throughout the other bars (opened in recent years) in New York, London, Hong Kong, Los Angeles, Porto Cervo and Miami.

Elements which have remained consistent to enhance the historical aspects of the brand include the original style décor, staff livery, standards of service, brand philosophy, table service, table and glassware and of course their infamous house cocktail – the Bellini.

Look through the brand history descriptions on the following websites:

http://www.harrysbarvenezia.com

http://www.cipriani.com/history.php

The brand elements required to build strong brands should be incorporated into a brand programme which manages the implementation of brand strategy and tactics.

2 Brand programmes

In Chapter 2 we looked at the elements of a brand. Clearly these need to be carefully planned and the process for doing this is to develop a brand programme.

2.1 The brand audit

The initial stage of the brand programme will be to conduct a brand audit to establish sources of brand equity from the perspective of both the firm and the consumer (Keller, 2008). This will enable the marketer to make informed strategic positioning decisions. The brand audit should identify:

- what consumers know about brands – what deeply held perceptions and beliefs create the true meaning of brands

- sources of brand equity and whether these are satisfactory

- from the firm's perspective – what products and services are currently offered

- whether any brand associations need to be strengthened

- whether the brand lacks uniqueness compared to competitors

- what brand opportunities exist

- the challenges to brand equity

According to Keller (2008) the brand audit should have two stages:

This stage provides a profile of how all the products and services offered are branded. The following items are included:

- brand name
- logos
- symbols
- characters or mascots
- packaging (or service environment style)
- slogans
- trademarks
- product or service attributes
- characteristics of the brand
- pricing
- communications
- third party (including celebrity) endorsements
- distribution policies
- digital brand elements
- other relevant marketing activities

Competitor brands should also be profiled to identify points of similarity and difference.

Activity 2

Complete a Brand Inventory for your own organisation.

Stage Two – Brand Exploratory

Actual consumer perceptions are investigated via the brand exploratory. Research is conducted to understand what consumers think and feel about the brands.

Existing research which may have already been conducted within the organisation provides a good starting point for the research.

Next, a range of internal employees should be interviewed to identify what they think the consumers' perceptions of the brands will be. You should also include their view of how your brand is deemed similar or different to competitors.

Finally primary research with consumers should be conducted. Both qualitative and quantitative methods are useful. The aim of qualitative research in brand audits is to probe beyond what consumers explicitly state to determine what they implicitly mean.

Quantitative research provides an overview of the depth and breadth of the brand and can be used to measure:

- brand awareness
- strength of favourability
- uniqueness of brand associations
- comparison against competitors
- perceptions of different aspects of the brand inventory eg logo, packaging, brand name etc

Keller (2008) has a highly detailed example of a brand audit conducted for Rolex at the end of Chapter 3.

2.2 Key programme decision areas

Some key decisions and activities in corporate or strategic branding will include:

- **Whether or not to brand**. Not all parent companies or manufacturers of product brands will want to be branded in their own right. Later we will look at the concept of umbrella branding. If synergies are not available from linking sub-brands, or associating them with a central corporate identity, the parent company may not be separately branded.

If you have never heard of Yum! Brands Inc, look up its web site: http://www.yum.com/company. You might be surprised to hear that it is the world's largest restaurant company, and even more surprised by its globally familiar, high-profile brands. Then click through to any of the brand sites, and see whether the Yum name or identity features on any of them

---

- **Selection and protection of brand name** (if not already determined as the company name). Desirable qualities for a brand name are (a) ease of pronunciation, recognition and recall and (b) resonance or suggestiveness (expressing something about the corporate identity).

- **Brand strategy and architecture**. Extent of integration or endorsement in the corporate brand in relation to subsidiary brands; and extent to which brand identity is internally and/or externally expressed.

- **Brand reputational platform**. The dimensions of stakeholder engagement with the brand: corporate activities (analogous to the features of a product brand); benefits or value added (analogous to the functional benefits of a product brand); and emotional benefits and associations (analogous to the emotional benefits of a product brand). The reputation platform provides a framework for which themes generate the greatest stakeholder engagement with the brand, and acts as a strategic communication framework for expressing the brand to target audiences. We look at brand architecture in more detail in section 3.

- **Brand promise / identity**. What the brand stands for or offers in the minds of stakeholders. A psychological 'contract' is formed between stakeholders and the brand, based on the values and benefits the brand holds out to its audience: the claims made and expectations raised by brand identity and associations. In the case of a bug killer, that means dead bugs when used. Easy to understand, easy to communicate. One famous branded bug killer, Raid® still used 'Kills bugs dead' as its advertising slogan. While it is easier to consider the branding challenge of a tangible product with hard and fast features, functions and benefits, the greater challenge is how does one communicate a corporate brand, especially if the corporate brand is really a holding company that owns many other brands?' (Doorley & Garcia, 2007 p. 283). What real, emotional or symbolic benefits can corporate brands offer or promise to stakeholders? This is again part of the brand's reputational platform.

- **Brand fingerprint** (Vyse, 1999). A document that summarises the essential character of the brand for everyone involved with the brand development process, to help ensure continuity and consistency of brand management. This may include:

    - **Target audience(s)**: description of the attitudes and values of the target audiences, in whose perception the corporate brand will be positioned. What attributes and values are relevant and important to them?

    - **Insight**: a description of the elements of stakeholder's perspectives and needs which will form a starting point for brand development

    - **Competition**: a picture of the market as seen by stakeholders, and the relative values the brand offers in the market

Developing Integrated Communications Strategy

- **Benefits**: the functional and emotional/symbolic benefits offered by the brand (brand promise)

- **Proposition**: the single most compelling and competitive statement the target stakeholder would make for engaging with the brand (purchase, investment, employment etc)

- **Values**: what the brand stands for and believes in

- **Reasons to believe**: the proof points offered to substantiate positioning

- **Essence**: the distillation of the brand identity into one clear thought (perhaps used as a corporate tagline or motto)

- **Properties**: the tangible and symbolic elements which will immediately evoke the brand.

2.3 Brand identity, promise and mantras

Brand identity: the message sent out by the brand through its product form, name, visual signs, advertising. This is not the same as **brand image** which is how the target market perceives the brand. The brand identity is developed as part of it's positioning. We discussed brand positioning in Chapter 2.

Brand mantras are frequently used to develop brand identity. Mantras are the articulation of the brand heart and soul (Keller, 2008). They are powerful devices which are often summarised in a few sentences or a short paragraph. They should suggest the core brand associations that consumers should hold. The Disney mantra for example is fun, family and entertainment. Nike's mantra is authentic, athletic and performance.

2.3.1 Three aspects to a brand

Aspect	Comment
Core	Fundamental, unchanging aspect of a brand. (Cider is an **alcoholic** drink made from apples.)
Style	This is the brand's culture, personality, the identity it conveys and so on. Compare the: • Rustic personality of Scrumpy Jack • Almost club-orientated personality of Diamond White
Themes	These are how the brand communicates through physical appearance of the product.

Clearly, the **themes** are more easy to change than the **style**, which is more easy to change than the core.

Brand reputation will be built according to the image and actions associated with the brand. This is linked to many of the brand elements as described above.

2.4 Brand reputation platforms

Definition

A reputation platform is the root positioning that a company or brand adopts when it presents itself to internal and external observers (van Riel & Fombrun, 2007 p. 136). Using a musical analogy (Hatch, 2003), it is the melodic 'riff' around which business unit managers improvise, interpreting and adapting the core message for the needs of different audiences.

A reputation platform creates a 'starting point' for the communication system: for all descriptions, discussions and expressions of organisational identity. Strong identities (and corporate brands) are built when logos, taglines, creative concepts and stories recognisably reflect an underlying reputation platform.

The core positioning themes that serve as reputation platforms may be:

- **What the organisation does**: the key activities or business areas it is involved in: eg DHL is in the logistics business; Shell is in the energy business. An activity theme emphasises focus, dedication, expertise and credibility in a given activity.

- **What the organisation offers**: the benefits or attractive outcomes offered by or through the organisation's activities. So Coles supermarkets are about 'everyday low pricing'; IBM is about 'working smarter, not harder'; 3M is about 'innovative and practical solutions from a diversified company'; Marks & Spencer is about 'trust, quality and service'. A benefits theme emphasises value, advantage and desirability, in order to inspire allegiance.

- **What the organisation represents**: the values and emotions the organisation stands for. So, for example, Reebok is about 'a passion for winning'; Volvo focuses on 'safety'; Johnson & Johnson on 'nurturing'; L'Oréal on 'you're worth it'; Intel on 'Today is so yesterday' (note the difference in tone from 3M's innovation theme, for example). An emotional theme seeks to establish a personal connection or emotional bond with target audiences, through empathy and identification.

2.5 Corporate stories and story-telling

> **Definition**
>
> A **corporate story** is 'a structured textual description that communicates the essence of the company to all stakeholders, helps strengthen the bonds that bind employees to the company, and successfully positions the company against rivals. It is built up by identifying the **unique elements** of the company, creating a plot that weaves them together, and **presenting** them in an appealing fashion' (van Riel & Fombrun, 2007 p. 144).

van Riel & Fombrun (2007) suggest that a good corporate story should be no longer than 400–600 words: a short but distinctive narrative about the organisation that (a) helps stakeholder audiences to understand it better and (b) distinguishes it from other entities in the reputation marketplace (*ibid*, p. 146). It will not necessarily be used for all audiences, or in its totality: it is intended primarily to guide corporate communication, as a briefing for media, marketing agencies, market analysts and other audiences who need to have an 'executive summary' of the essence of the organisation. Expansions, interpretations and versions of the story may, however, be used widely within and beyond the firm.

Unique corporate features may not be easy to identify, given the fairly narrow range of values likely to appeal to stakeholder audiences. (Doesn't every company want to represent customer focus, innovation, quality, value, responsibility, trustworthiness etc?) However, the attributes identified as central, consistent and distinctive, as part of corporate identity analysis, will provide a good starting point. In addition, each organisation will have unique historical roots and key personnel, which may form part of its identity.

There is no substitute for browsing corporate web sites (especially those which reflect major corporate brands) to appreciate the power of reputation platforms and corporate stories. For example, check out:

- 'Finding Better Ways' (A Brief Introduction to 3M), downloadable from:
  Link: http://solutions.3m.co.uk/wps/portal/3M/en_GB/about2/Our-Company

  See also the fascinating historical narrative at:
  Link: http://solutions.3m.co.uk/wps/portal/3M/en_GB/about-3M/information/more-info/history

- The timeline of Marks & Spencer's history at:
  Link: http://corporate.marksandspencer.com

  See also the material offered on the 'Student Information' link, and the new reputation platform represented by Plan A (M & S's CSR platform):
  Link: http://plana.marksandspencer.com

- The corporate responsibility platform of McDonalds:
  Link: http://www.mcdonalds.com/usa/good/welcome

- The 'Company overview' of Nike:
  Link: http://www.nikebiz.com/company_overview

- The 'Ford Story' ('See where we are. Be part of where we're going' – including blogs and social networking site links)
  Link: http://www.ford.com/about-ford/company

Even if features are not particularly unique, they can be giving distinctive meaning when they are connected and expressed in a **narrative** or **plot**. A story may be a simple narrative history, or statement of what the company stands for and believes in; what it is trying to do; or how it sees its future. However, it can also be made more vivid, in order to create powerful resonances with archetypal narrative themes. In his book *Seven Basic Plots*, Christopher Booker (2004) argues that there are seven basic plot lines: Comedy (confusion resolved in a happy ending), Tragedy (forces driving towards an unhappy ending), Rags to Riches, Journey and Return, The Hero's Quest (searching for something of great value), Overcoming the Monster, and Rebirth. You may be able to imagine how these stories would play out for an organisation: fighting off takeover, natural disaster or recession (overcoming the monster); persevering through obstacles to arrive at an innovative breakthrough (quest); starting small and becoming a global corporation (rags to riches); diversifying and then returning to its roots (journey and return); or undergoing a paradigm shift and turning the business around (rebirth).

If this seems fanciful, consider the 'years of struggle' of 3M before it found its innovation breakthrough. Or the 'rise and fall and rise' of Marks & Spencer in the last few decades – and now its heroic mission to save the planet (*Plan A will see us working with our customers and our suppliers to combat climate change, reduce waste, safeguard natural resources, trade ethically and build a healthier nation. We're doing this because it's what you want us to do. It's also the right thing to do. We're calling it Plan A because we believe it's now the only way to do business. There is no Plan B.'*). Consider the rags to riches story of Virgin 'beginning in the 1970s with a student magazine and small mail order record company' (www.virgin.com).

Effective corporate stories (van Riel & Fombrun, 2007 p. 146):

- Introduce **unique words** to describe the organisation
- Refer to the organisation's **unique history**
- Describe the organisation's **core strengths**
- **Personalise and humanise** the organisation
- Provide a **plot line**
- Address the concerns of **multiple stakeholders**

3 Brand architecture

Brand architecture is often taken to mean:

- The brands / sub brands and endorsed brands
- The roles of each of the brands and sub-brands

Aaker (2004) renamed brand architecture calling it instead brand portfolio strategy. He says that

"the brand portfolio strategy specifies the structure of the brand portfolio and the scope, roles, and interrelationships of the portfolio brands" p. 13.

3.1 Typologies of branding strategy

There are several basic branding strategies: company, umbrella, range and individual product.

3.1.1 Different types of brand strategy

(a)   **Company brand**. The company name is the most prominent feature of the branding (eg Mercedes).

(b)   **The company brand combined with an individual brand name** (eg Kellogg's: Corn Flakes, Rice Krispies). This option both legitimises (because of the company name) and individualises (the individual product name). It allows **new names to be introduced quickly and relatively cheaply**. Sometimes known as **umbrella branding**, firms might use this approach as a short-term way to save money.

(c)   **Range brand**. Firms group types of product under different brands. For example, Sharwoods is a brand owned by RHM Foods. Sharwoods offers pickles, poppadums, sauces etc.

(d)   **Individual name**. Each product has a unique name. This is the option chosen by Procter & Gamble for example, who even have different brand names within the same product line, eg Bold, Tide. The main advantage of individual product branding is that an unsuccessful brand does not adversely affect the firm's other products, nor the firm's reputation generally.

Penguin is one of the oldest brands in UK paperback publishing and over the years has introduced brand extensions (*Puffin* for children, *Pelican* for academic) and sub-brands (*Penguin Classics, Penguin Modern Classics*) and indeed other products (*Penguin Classic CDs*).

A key issue for publishing is to identify the core of the brand.

(a)   The imprint or publisher?

(b)   The author? It appears to go without saying that people will buy a book by a recognised author, and that the author is at the heart of the brand.

In contrast, people buy '*Mills & Boon*' books – the core of the brand is the publisher, not the author.

The *Folio Society* publishes versions of classic literature, but markets its books partly as art objects, owing to the quality of the binding and paper, and the specially-commissioned illustrations.

3.1.2 Olins

Olins (1990) proposed an influential typology of corporate branding strategies.

- **Monolithic strategy**: the whole company uses one visual style and symbols, for instant recognition. This suits companies which have developed as integrated entities within a relatively narrow field (eg Shell, Philips, BMW).

- **Endorsed strategy**: different subsidiaries or divisions have their own style, but the parent company is clearly identified and recognisable. This suits diversified companies whose divisions have distinctive culture and brands
(eg General Motors, L'Oréal, Altria, Kelloggs).

- **Branded strategy**: different subsidiaries or divisions have their own style, without visible relationship to each other or to the parent, which is a company more or less invisible (eg Unilever).

Monolithic and endorsed strategies enable (positive or negative) 'image spillover' or 'halo' effects to different degrees. A branded strategy does not allow the brand to benefit from the parent company's positive reputation – but it also limits the risk to the corporate brand of product failure or reputational damage.

The following working paper from Judge Business School, University of Cambridge looks at the branding strategy and the likely effect on corporate financial performance:
http://www.jbs.cam.ac.uk/research/working_papers/2006/wp0613.pdf

3.1.3 Kammerer – corporate branding

Kammerer (1988) categorises implementation of corporate branding strategies by four 'action types':

- **Financial orientation**. Subsidiaries are purely financial participants in the overall concern: they are more or less autonomous of the parent company, and manage their own identities.

- **Organisation-oriented corporate branding**. The parent company exercises influence over the culture and strategy of the subsidiaries (eg by setting policy and rules), but this is a form of internal corporate branding – not directly visible to the outside world.

- **Communication-oriented corporate branding**. Advertising and visual identity clearly express the fact that the subsidiaries are part of the parent company – supporting added value from endorsement (as discussed earlier).

- **Single company corporate branding**: similar to monolithic branding, with integration of action, messages and symbols across the whole concern.

3.1.4 van Riel and van Bruggen corporate branding typology

The model developed by van Riel and van Bruggen (2002) takes into account two key factors in the corporate branding decision, which explicitly acknowledge the influence of business unit managers on the successful implementation of corporate brands:

- **Agreement on parent visibility**: the extent to which business unit managers are willing to communicate that they are part of a larger group of companies

- **Agreement on starting points**: the extent to which there is consensus about the starting points of the corporate branding strategy (what the parent company really stands for, what its values are, and how they can be used to communicate with target audiences).

These two dimensions create four basic choices: van Riel and van Bruggen's (2007) *corporate branding typology*

<table>
<tr><td rowspan="2">Agreement on
parent visibility</td><td>*High*</td><td>Medium endorsement</td><td>Strong endorsement</td></tr>
<tr><td>*Low*</td><td>No endorsement (stand alone)</td><td>Weak endorsement</td></tr>
<tr><td></td><td></td><td>*Low*</td><td>*High*</td></tr>
<tr><td></td><td></td><td colspan="2">Agree on starting points</td></tr>
</table>

Examining each quadrant in turn:

- **No endorsement** (stand alone). There is a high degree of autonomy in business units, and a low degree of parent visibility, in order to avoid image spillover effects. Visualisation: affiliate/subsidiary name (eg Lipton).

- **Weak endorsement**. Usually a transitional phase, allowing low parent visibility while consensus and support is built for a more integrated market approach. Visualisation: affiliate name + 'member of' parent company name + parent company logo (eg Lipton a member of Unilever [logo]).

- **Medium endorsement**. There is a high degree of parent visibility, but no consistent integration with corporate message. Visualisation: parent company name (logo) + affiliate/subsidiary name (eg Unilever [logo] Lipton). Designed to bolster the strength of the affiliate brand.

- **Strong endorsement**. There is a high degree of parent visibility and corporate identification, transparency, co-ordination of communication strategies: designed to show the strength of the group as a whole. Visualisation: parent company name (logo) + specialisation (eg Unilever [logo] Food and Beverage). (If you are interested in how Unilever actually approaches this, see: http://www.unilever.com.)

Definition

Endorsement (in the context of corporate branding strategy) is the extent to which a subsidiary brand is associated with, or given approval or sanction by, a parent brand: in other words, the degree of parent company visibility in the management and visual identity of the subsidiary brand.

A business unit should only move towards a stronger degree of endorsement as:

- The corporate brand grows more well-known and valued in the local market
- The local brand loses strength in its local market, as the importance of the corporate brand grows.

In practice, this may mean that different business units should be assessed on a case by case basis.

The **Ford Motor Company** is a strong 'family of brands'. The Ford name is a brand family name for a number of separately branded models (Falcon, Fiesta and so on). It has also been stretched to cover other activities: for example, the Ford Motor Credit Co (offering automotive finance for dealers and customers of other Ford corporate brands).

However, after a series of acquisitions, Ford Motor Company also controls a range of corporate names – strongly branded in their own right – such as Volvo, Mercury and Lincoln. In the dealership environment, Ford may locate their different marques on the same, or adjacent sites, but they keep the experience separate: Volvo drivers are, looking for a different product and service experience than that expected by a Ford customer. Similarly, while the Ford Motor Company web site and advertising emphasises the 'Family of Brands' and features Ford, Mercury, Volvo and Lincoln side by side, each subsidiary brand has its own distinctive visual identity and corporate brand. (The Volvo web site does not identify it as a member of Ford, for example.)

Link: http://www.ford.com/about-ford/company-information
http://www.volvocars.com

Identify the brand architechture in operation within your own organisation.

3.2 Choice of brand strategy

(a) **Company and/or umbrella brand name**

Advantages	Disadvantages
• Cheap (only one marketing effort)	• Not ideal for segmentation
• Easy to launch new products under umbrella brand	• Harder to obtain distinct identity
• Good for internal marketing	• Risk that failure in one area can damage the brand
	• Variable quality

For example, Virgin is a company brand name, supported by advertising, PR and the celebrity status of Richard Branson. (To what extent will the problems of Virgin Trains adversely affect the other brands?)

(i) Service industries use umbrella marketing as customer benefits can cross product categories. Marks & Spencer diversified from clothes, to food and to financial services. Tesco and ASDA have followed suit.

(ii) Communication media are more diffuse and fragmented.

(iii) One brand is supported by integrated marketing communications.

(iv) Umbrella branding supports **database marketing** across the whole product range.

(v) Distributor/retailer brands are umbrella brands in their own right, so **brand owners** have to follow suit.

(b) **Range brands** offer some of the advantages of an umbrella brand with more precise targeting.

(c) **Individual brand** name

Advantages	Disadvantages
• Ideal for precise segmentation	• Expensive
• Crowds out competition by offering more choice	• Risky
• Damage limitation to company's reputation	

Factory equipment brands

MG Technologies industrial group, the company that owns the Tuchenhagen brand, promotes the name as part of a multi-brand philosophy. At Keyence, one of the world's biggest producers of sensors and vision systems for factory processes, the brand management is somewhat simpler; the Keyence name, rather than specialist 'sub-brands', is the brand most heavily promoted by the company.

Another leader in running different brands within the same business is Sandvik, the world's biggest manufacturer of machine-tool devices.

Sandvik's tooling division has about 10 key brands. They include Coromant, which is associated with particularly hard cutting materials; Valenite, aimed at automotive applications; Walter (general machining); and Titex (drilling). Most of the company's advertising and marketing effort is aimed at establishing the value of these brands, rather than raising awareness of the Sandvik name itself.

Assessment advice

When you are auditing your own brand, consider the elements covered within this chapter and think particularly about issues such as the brand architecture in place and whether you would look to change this.

1 Identify the steps to build strong brands

- Strong brands are those where marketers have a good appreciation of the asset types constituting their brands
- Consistent values are essential

2 Coordinate brand programmes

- Brand audits should direct the overall brand strategy.
- Consist of a number of elements including the brand strategy or architecture, promise and brand fingerprint.

3 Design brand architecture

- The strategy designed for the brand.
- A range of typologies to explain brand strategy exist.
- There are several basic branding strategies: company, umbrella, range and individual product

1   Working through Keller's ten commandments may have taken you quite a while. It is not an easy activity and may be worth reviewing with a colleague or possibly your programme Study Buddy.

2   You will be able to use this activity to assist you with your assignment prep.

3   This will depend on your own research but will be useful to your assignment prep.

Aaker, D.A., (1996). *Building Strong Brands*. New York: The Free Press.

Booker, C., (2004). *Seven Basic Plots: Why We Tell Stories*. Continuum International Publishing Group

de Chernatony, L. and McDonald, M., (2003). *Creating Powerful Brands*, Oxford: Butterworth-Heinnemann.

Hatch, M. J. & Schultz, M., (2000). 'Scaling the Tower of Babel' in Schultz, Hatch & Larsen (eds) *The Expressive Organisation*. New York: OUP pp. 11-35.

Keller, K., (2008). *Strategic Brand Management*. International edition. 3rd edition. Harlow: Prentice Hall.

Olins, W., (1990). *The Wolff Olins Guide to Corporate Identity*. London: The Design Council.

van Riel, C. B. M. & van Bruggen, G. H., (2002). 'Incorporating Business Unit Managers' Perspectives in Corporate Branding Strategy Decision Making' in *Corporate Reputation Review* Vol 5 (2/3) pp. 241-251.

van Riel, C. and & Fombrun, C. J., (2007). *Essentials of Corporate Communication: Implementing Practices for Effective Reputation Management*. Abingdon, Oxon: Routledge.

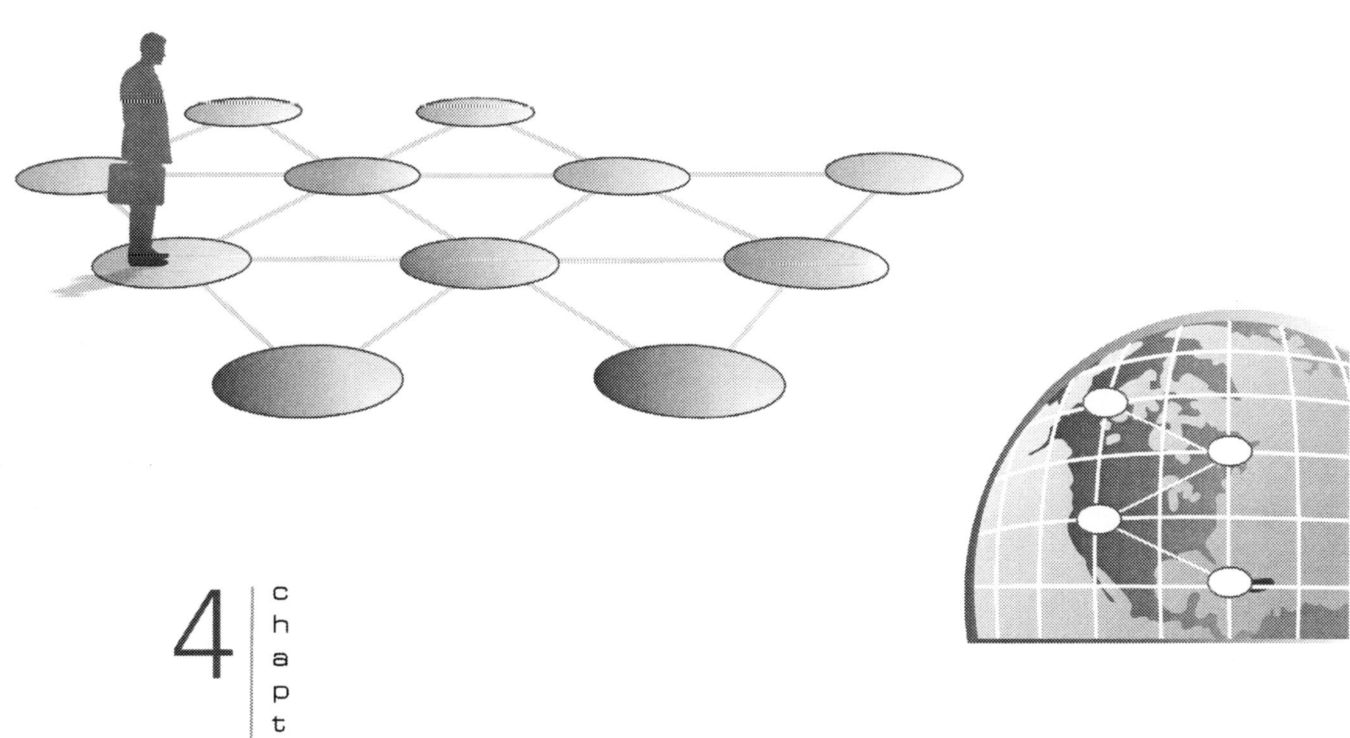

4 | chapter

Managing brands

The importance of the financial value of brands has been highly significant for organisations since they have been placing brand values on their balance sheets.

The monitoring and control of brands is not only required for financial analysis however but it is vital to ensure that brands are managed successfully over time and corrective action is taken where necessary.

Over time, brands often need to extend to remain fresh while maintaining the consistency of their core values and associations. Brand planning is critical especially at this stage. Extensions come from either diversifying into new product areas of entering new markets.

When looking at new markets, global issues require consideration and the chapter covers the nature of global branding.

Contents

By the end of this chapter you will be able to:

- Discuss how to manage brands over time
- Suggest alternative methods of measuring and valuing brands
- Identify the complexities of global branding decisions

1 Managing brands over time

As brands progress through their product life cycle, brand reinforcing activities need to be considered. At the introductory stage the brand requires building but over time brand extensions and refreshing activities are required.

1.1 How to build brands

Brand building is a logical process that can proceed gradually, step by step towards a position of strength. In Chapter 3 we looked at what constituted a strong brand and the elements to be included in brand strategy. Brand audits were discussed in Chapter 3 as the starting point for developing strategy.

The process for building the brand is similar to that of building a product (core product, an expected product, an augmented product and a potential product). However, a product is, in some respects, purely functional, whereas a brand offers more. We discussed brand programmes, brand strategy and architecture in Chapter 3. Remember these concepts as you work through the rest of this chapter.

1.2 A step approach to designing brands

Step 1	Have a quality product – but remember quality means fitness for use not the maximum specification. Functionality is only a starting point.
Step 2	Build the basic brand. These are the marketing mix criteria.

- They should support product performance
- They should differentiate the brand
- They should be consistent with positioning
- The basic brand delivers the core product in an attractive way

Step 3	Augmentations include extra services, guarantees and so on. (Expensive guarantees provide evidence that the firm takes quality seriously.)
Step 4	Reaching its potential, so that customers will not easily accept substitutes.
Step 5	Maintain brand value by using the marketing mix to persuade customers to re-buy.
Step 6	Build brand loyalty. Customers who rebuy and are loyal are valuable because:

- Revenue from them is more predictable
- Existing customers are cheaper than new customers

Step 7	Know where to stop in developing the brand. (For example, an alcohol-free alcopop would be pointless because it would just be a soft drink – so may as well be marketed as an alternative brand.)

Brands that **reach their potential over time** have five key characteristics.

- A quality product underpinning the brand
- Being first to market, giving early mover advantages
- Unique positioning concept: in other words they are precisely positioned
- Strong communications underpinning the brand
- **Time and consistency**

1.3 The brand planning process

Over time the brand programme is likely to need re-tuning. Brand strategy is one of the steps in the brand planning process just as marketing strategy is one step in the marketing planning process. Arnold (1992) offers a five stage brand planning process.

Stage	Description
Market analysis	An overview of trends in the macro and micro environment and so includes customer and competitor analysis and the identification of any PEST factors which may affect our brand. For soft drinks, the explosion of competitive activity, particularly by own label, and new product introductions, such as Fruitopia, will be important.
Brand situation analysis	Analysis of the brand's personality and individual attributes. This represents the internal audit and questions such as, 'Is advertising projecting the right image?', 'Is the packaging too aggressive?', 'Does the product need updating?' need asking. This is a fundamental evaluation of the brand's character.
Targeting future positions	This is the core of brand strategy. Any brand strategy could incorporate what has been learnt in steps (1) and (2) into a view of how the market will evolve and what strategic response is most appropriate. Brand strategy can be considered under three headings. (1) Target markets (2) Brand positions (3) Brand scope
Testing new offers	Once the strategy has been decided the next step is to develop individual elements of the marketing mix and test the brand concept for clarity, credibility and competitiveness with the target market.
Planning and evaluating performance	The setting of the brand budget, establishing the type of support activity needed and measurement of results against objectives. Information on tracking of performance feeds into step (1) of the brand management process.

Global case study

Sunny Delight

This is an example of what can go well – and – not well in planning, introducing and managing a new product. (Extracts from *The Guardian*, 11 April 2001.)

Sunny Delight burst upon Britain with its sunshine logo in April 1998. By August 1999, it was the country's third-largest-selling soft drink. Three years later, sales were down 36% by value and 28% by volume (moving annual totals to February 2001).

It was a textbook launch. Delight had been available in the US since 1964, it was sold as a downmarket drink competing for space alongside squashes and long-life drinks on ordinary shelves. The approach in the UK was to be different. Procter and Gamble (P&G), one of the world's most powerful grocery manufacturers, had acquired it at the end of the 80s and in 1996 began a long and thorough process of test marketing it for the UK in Carlisle.

Delight is 5 per cent citrus juice, and a lot of sugar and water, with vegetable oil, thickeners, added vitamins and flavourings, colourings and other additives that make it look like fresh orange juice but appeal to the immature tastebuds of young children.

The ingredients were cheap but the price was set at a premium. P&G invested in a new filling plant costing about £12m, according to industry estimates, so that the drink could be packaged in the sort of frosted plastic bottles that fresh orange juice is usually sold in chill cabinets, next to fresh fruit juices.

P&G is one of the handful of companies that has the muscle to dictate where products are sold in supermarkets. All this was backed up by a huge direct marketing campaign and a £9.2m advertising campaign.

P&G's brands include Pampers, and it is thought to have built up a powerful database from offers over the years, which tells the company who we are and how old our children are ... it is also reported to have worked with retailers' data from loyalty cards to identify young, lower-income families. Teenagers were targeted with sponsorship of basketball.

This combined onslaught led to instant success. But the backlash came equally fast. The Food Commission condemned Sunny Delight as a con, accusing P&G of putting it in chill cabinets to mislead. Newspapers, the BBC's *Watchdog* programme and *Radio 4* all carried attacks on the brand and dubbed it 'The unreal thing'.

Then came the comic twist in the drama. In December 1999, a paediatrician, Dr Duncan Cameron, reported a new and alarming condition in the medical journals: Sunny Delight syndrome. A girl of five had turned bright yellow and orange after drinking 1.5 litres of the stuff a day. She was overdosing on betacarotene, the additive that gives the sugar-and-water drink its orange colour.

By a marketing man's nightmare of coincidence, the TV ads for the brand at the time showed two white snowmen raiding the fridge for Sunny Delight and turning bright orange. To add to the embarrassment, a leading consultant dermatologist, Professor John Hawks, said too much betacarotene could cause tummy upsets. As if to confirm its status as spawn of the devil, P&G was forced to join that happy band of cigarette manufacturers who put voluntary warnings on their products.

1.4 Brand extension

A brand can be used on a wide range of products and services if its values are appropriate.

Brand extension uses a brand name successfully established in one market or channel to enter another. It is often termed **brand stretching** when the markets are very different.

1.4.1 Examples of brand extension

- Retailers such as Tesco launching themselves as price comparison service and financial services provider etc

- easyJet launching easyCruise, easyHotel, easyCar

- KitKat extending from one to over 20 product formats

- Coppertone sun lotion launching sunglasses

- Designer Kath Kidston licensing textile designs to be used for re-usable Tesco carrier bags, Roberts digital radios, mobile phones and a range of other products

- One of the greatest exponents of brand extension is Richard Branson. He has extended the Virgin brand, originally based on pop music, to cover mobile phones, trains, soft drinks, weddings, car hire, financial services and even space travel!

The following website provide useful case studies of brand extensions:
http://www.brandextension.org/

1.4.2 Conditions for brand extension

(a)   The **core values** of the brand must be **relevant to the new market**. EasyJet has transferred to car rental and internet cafes.

(b)   The new market area must not affect the core values of the brand by association. Failure in one activity can adversely affect brand equity.

Consider how your brands might be extended.

Mark Ritson, assistant professor of marketing at the London Business School, wrote about two planned attempts at brand stretching in *Marketing* (April 2004). He did not expect either to succeed.

- Stelios Haji–Ioannou, founder of easyJet has announced easy4men, a range of male grooming products. Ritson doubts this will succeed as easyGroup does not really have 'brand equity in the form of positive, valuable and extendable brand associations', it merely has 'an unusual business model built on a stripped-down product offering and dynamic pricing.'

- *The Daily Telegraph* is to launch a compact edition. Ritson expects this to fail because the Telegraph 'is tradition, it is conservatism, it is quality. It is all the things that a compact edition is not.' He contrasts the Telegraph's prospects with the successful launch of a compact edition of *The Independent*, 'a younger, different and more contemporary newspaper brand.'

easyGroup's partnership with Boots was dissolved in 2006, though a single easy4men product (a 'three-day travel pack') remained available. The compact edition of *The Daily Telegraph* seems to have disappeared without trace.

1.4.3 Advantages of brand extension

Advantage	Comment
Cheap	It is less costly to extend a brand then to establish a new one from scratch.
Customer-perception	Customer expectations of the brand have been built up, so this lower risk for the customer encourages 'trial'.
Less risky	Failure rate of completely new brands.

1.4.4 Disadvantages of brand extension

Advantage	Comment
Segments	The brand personality may not carry over successfully to the new segment. The brand values may not be relevant to the new market.
Strength	The brand needs to be strong already.
Perception	The brand still needs a differential advantage over competitors.
Over-dilution	Excessive extensions can dilute the values of the brand.

1.5 Revitalising

At times, the performance of a brand will falter and managers will attempt to rectify the situation by enhancing sales volume and improving profits in other ways.

Revitalisation means increasing the sales volume through:

- New markets (eg overseas)
- New segments (eg personal computers are being sold for family, as opposed to business, use)
- Increased usage (encouraging people to eat breakfast cereals as a snack during the day)

Repositioning is more fundamental, in that it is a **competitive strategy** aimed to change position to increase market share.

Type of position	Comment
Real	Relates to actual product features and design
Psychological	Change the buyer's beliefs about the brand
Competitive	Alter beliefs about competing brands
Change emphasis	The emphasis in the advertising can change over time

1.6 Success criteria for branding

Beneficial qualities of a brand name

- Suggest **benefits**, eg Schweppes' Slimline tonic
- Suggest qualities such as **action** or **colour** (eg easyJet, with an orange colour)
- Be **easy to pronounce**, recognise and remember
- Be **acceptable in all markets**, both linguistically and culturally
- Be **distinctive**
- Be **meaningful**

Global case study

Compare the following mobile phone brands.

- Orange
- O_2
- Vodafone

Orange appropriates the colour orange, whereas Vodafone and O_2 suggest aspects of the product. Review the brands websites to identify the differences for yourself.

www.orange.co.uk
www.O2.co.uk
www.vodaphone.co.uk

2 Brand measurement and valuation

Keller (2008) notes:

"Ideally to measure brand equity, we would create a 'brand equity index' – one easily calculated number that would summarize the health of the brand and completely capture its brand equity. But just as a thermometer measuring body temperatures provides only one indication of how healthy the person is, so does any one measure of brand equity provide only one indication of the health of the brand"

As such it is not sufficient to take one measure at one point in time to accurately calculate the value of brands.

Both qualitative and quantitative techniques should be used to measure brands as we have already discussed in terms of conducting a brand audit.

The following methods are suitable:

Qualitative brand measures	Quantitative brand measures
Free association	Brand awareness – prompted and unprompted
Adjective ratings and checklists	Direct and indirect brand recognition
Projective techniques	Brand image:
Photo sorts	• scale measures of brand attributes
Bubble drawings	• strength, favourable, uniqueness
Story telling	• overall judgements
Personification	• relationship measures (intensity and frequency)
Role playing	
Experiential methods	

In terms of actual metrics applied, the following table summarises

Method	Application	Critique
Brand based comparison Competitive brands used as a benchmark	Blind testing Testing within the home or usage setting Qualitative comparisons between brands	Positive aspects – it isolates the value of the brand. Useful in developing strategies for advertising, pricing as reacted to by consumers and a wide range of marketing activities. Disadvantages can be that by being used as a research item, the brand values and salience may be artificially highlighted to the consumers and so results are distorted.
Marketing-based comparison	Based on demand curves to assess price sensitivity thresholds for brands eg what premium would consumers be willing to pay in different circumstances. Also used to assess response to advertising strategies, media plans, copy testing, and multiple test markets.	This is an easy to implement measure. Virtually any proposed set of marketing actions can be compared. The main drawback is that responses cannot necessarily be attributed to the brand values. It may be a generic product behaviour in force.

Method	Application	Critique
Conjoint analysis	Survey based multivariate technique which enables marketers to profile consumer decision processes for brands. Sometimes this method is simplified with just two variables such as brand and price. Consumers would make a series of simulated purchase choices between different combinations of brands and prices.	Ability to gain knowledge about buying profile of consumers in relation to a specific brand. However this is often a complex methodology where brands are difficult to interpret brand attribute levels.

2.1 Brand valuation methods

There are a number of both generic and proprietor methodologies available to value brands. Keller (2008) termed the major methods as 'holistic' methods because they place an overall value on the brand in either abstract or concrete financial terms.

The two main holistic approaches are:

- **Residual approach** – value of the brand based on subtracting consumers' preferences for the brand based on the physical product attributes alone. These provide a useful benchmark for interpreting brand equity but they are not appropriate for brands with many functional related brand associations because they cannot distinguish functional from non-functional attributes.

- **Valuation approach** – places a financial value on brand equity for accounting purposes, mergers and acquisitions.

Separating out the percentage of revenue or profits attributable to brand equity is a difficult task. Three main approaches are used:

- **Cost method** – brand equity is the amount of money that would be required to reproduce or replace the brand including R&D, test marketing, advertising etc

- **Market method** – brand equity is the present value of the future economic benefits to be derived by the owner of the asset.

- **Income approaches** – brand equity is the discounted future cash flow from the future earnings stream for the brand (eg capitilising royalty earnings from the brand name, premium profits earned by a branded product or capitalising the actual profitability of a brand after allowing for the costs of maintaining it and taxation).

GMN viewpoint

GMN Advisory Council member David Haigh is the founder of Brand Finance. Brand Finance is a leading global consultancy specialising in the valuation of brands.

The consultancy helps clients' value, articulate and build their intangible asset base using language and approaches understood by financial, marketing and investor audiences.

The proprietal techniques used are outlined in the following paper availible to download:

http://www.brandfinance.com/Uploads/pdfs/CurrentPracticein_Brand_Valuation.pdf

A number of proprietor techniques exist within consultancies who specilise in brand valuation. Often these incorporate marketing, financial and legal aspects using detailed research and following fundamental accounting concepts.

Some of the best known agencies include:

Brand Finance – www.brandfinance.com

Interbrand – www.interbrand.com

Young and Rubicam's BrandAsset Valuator – www.yrbav.com/

Assessment advice

You are not expected to value your brand within your assignment but you are expected to know which metrics you could use to measure your brand strategy.

Which of the methods for brand valuation we have discussed in this section do you think is more appropriate to use in your own organisation?

3 Global branding

3.1 What are global brands?

It is hard to build truly global brands, in other words a brand whose positioning, advertising strategy, personality, look and feel are in most respects the same from one country to another. There are examples of success, for example Visa and Mastercard in financial services, Coca-Cola in soft drinks and so on.

According to Kotabe and Helsen (2004) a *truly* global brand:

"has a consistent identity with consumers across the world. This means the same product formulation, the same core benefits and values, the same positioning. Very few brands meet these strict criteria."
p. 354

GMN viewpoint

GMN Faculty Member Svend Hollensen suggests that ALL brands compete on a global scale. Even if you are not currently selling on a global scale you are still competing with overseas competitors who may capture some of your maket share. Awareness of global issues are then essential for ALL marketers.

Consider the case of airlines. Is Virgin a global brand? Is Singapore Airlines global? After all, anyone taking an international flight from Singapore to another destination is likely to experience the same core benefit (air transportation), physical evidence (aircraft cabin). Microsoft software is installed on many of the world's PCs, in businesses and in the home. Is it a global brand as described above?

Intel chips feature in many adverts with a distinctive 'jingle' (tune). Is Intel a global brand for computer manufacturers and/or those who buy computers with Intel micro-processors?

Accenture consulting offers consulting services to businesses all over the world. (Accenture used to be Andersen Consulting, but changed its name – some time before Arthur Andersen was wound up.)

3.2 Brands in the global market place

Brands, like most aspects of marketing practice, require a decision about standardisation or adaptation when taken into the global marketplace.

The most successful examples of worldwide branding occur where the brand has become **synonymous with the generic product** (eg Sellotape, Aspirin, Windsurfer).

Brewing is an industry with significant economies of scale. Apart from Heineken and Guinness, it is only recently that big brewers have become 'international'. There are a variety of aspects of this development.

(a) Beers are branded across markets. Stella Artois is available in the UK as a premium product, whereas in Belgium it is 'a decent modestly price lager'.

(b) Other firms are expanding by acquisition. Interbrew, the brewer of Stella Artois has purchased Labatt of Canada, to gain access to markets in North and South America.

(c) Big brewing companies see many European and American markets are stagnant: they are trying to revive them with imported or foreign brands.

(d) Firms co-operate in some markets but compete in others. (Guinness distributes Bass in the US, whilst competing with Bass in the UK.)

(e) The greatest potential seems to be east Asia, where beer consumption is rising by 10% pa and South America, where growth is 4% pa.

(f) Cobra is branded as an Indian beer but is actually a British brand originally distributed only through Indian restaurants within the UK. A recent brand repositioning has widened the usage opportunities to highlight the lower gas content of the beer making it more drinkable in a wider variety of social occasions. More varied distribution channels are utilised to reinforce this value.

Ultimately, even if the beer market eventually becomes truly global, it will remain fragmented for a long time to come.

Developing Integrated Communications Strategy

Blue Nun

Blue Nun was relaunched around the globe as an affordable, reliable and widely available bottle of wine. Having been in steady decline in the UK since 1985, sales recovered with half a million cases sold in 2000. The target market in the UK is women over 35 who want to spend around £5 per bottle, while in Asia the core target market is slightly younger as they are newcomers to the wine market.

The wine is sold in over 80 countries. There is a 'brand book' with strict guidelines for local distributors on such matters as brand values and which fonts and colours to use. This should ensure consistency of brand message.

Blue Nun was one of the first more modern 'branded' wines. Today, with the global popularity of New World, California, Chilean and other regional wines, branding has taken a more active role.

French wines have been hit by the trend since around 2005 as they were not keen to employ the more modern integrated branded approaches and relied instead on the one dimensional aspect of their labelling.

Budweiser is a US beer, but a beer with an identical name (though a very different taste) is made in the Czech Republic and is sold throughout Europe. Anheuser-Busch, the American owner, has been unable to buy the Czech beer's trademark. In 2007, it was reported that the two companies had agreed to collaborate on distribution in the USA.

3.2.1 Cultural aspects

Even if a firm has no legal difficulties with branding globally, there may be cultural problems, eg unpronounceable names or names with other meanings. There are many examples of problems in global branding, for example 'Maxwell House' is 'Maxwell Kaffee' in Germany, 'Legal' in France and 'Monky' in Spain. But sometimes a minor spelling change is all that is needed, such as 'Wrigley Speermint' in Germany.

3.2.2 Other considerations

Many other influences affect the global branding decision.

(a)   Differences between the firm's major brand and its secondary brands. The major brand is more likely to be branded globally than secondary brands.

(b)   The importance of brand to the product sale. Where price, for example, is a more important factor, then it may not be worth the heavy expenditure needed to establish and maintain a global brand in each country; a series of national brands may be more effective.

(c)   The problem of how to brand a product arising from acquisition or joint venture. Should the multinational company keep the name it has acquired?

Questions in this area often will specifically ask for examples, so you should keep your eyes on the business press. The draft specimen paper included a 25-mark question on developing a global brand and asked for examples of the way organisations have applied critical success factors in brand development.

3.3 The benefits of global branding

A global brand can offer some benefits to a company.

Benefit	Yes, but ...
Economies of scale The development costs of the products and other areas of the mix can be carried across a global market.	Economies of scale can prove elusive, especially if subtle product adaptations need to be made. Locally produced ads may be cheaper, and more effective, than imports.
Visibility The brand is visible all over the world, and so may be exposed to more consumers (eg tourists)	The important thing, though, is positioning. In Japan, cars are built to a high manufacturing quality, so a positioning on a 'quality' basis is not appropriate. In countries where there are differences in reliability, a quality positioning can be of value. The brand might 'mean' different things in overseas markets.
Prestige	A brand may get prestige by its global positioning. (British Airways once rebranded itself as the 'world's favourite airline').

3.4 Differences in brand equity across different country markets

Although some brands may be available world wide, this does not mean that they are salient in all markets. The value of a brand, its brand equity, often varies significantly from one country to the next. These differences can be the result of any of the factors below:

History	Well-established brands that have existed for a long time have more familiarity among consumers than brands that were introduced more recently. Brand image is likely to be stronger when a consistent positioning strategy has been used over the years.
Competition	The more intense the competition in a given market, the more difficult it becomes for brands to stand out from the rest.
Marketing support	This can vary from country to country, particularly when organisations have a decentralised structure. Some subsidiaries may be in favour of trade promotions and other incentives geared towards distributors, while others might prefer to focus their marketing on the end consumer. This can create differing levels of brand equity. The power of distributors in a market can affect the type of approach used. In the UK, for example, retailers are very powerful and often demand significant portions of manufacturers' marketing budgets to be spent on sales promotions and other activities; this is not necessarily the case elsewhere.

Cultural receptivity	If a brand denotes membership of a group, cultures with a **collectivist** mentality will favour branding. If brands are a sign of quality, then risk aversion will favour branding. In **individualist** cultures, a brand could be seen as a sign of status, and hence something to aspire to. On the other hand, an individualist could reject branding as it signifies mindless conformity to someone else's agenda.
Other products in the category	If the type of product is widely used, a strong brand will have more salience. Dove is a strong brand as washing and shower products are widely purchased.

As well as Ten emotional branding commandments (see Chapter 3), Keller (2009) has provided "Ten Commandments of Global Branding".

	Commandment
1	Understand similarities and differences in the global branding landscape
2	Don't take short cuts in brand building
3	Establish marketing infrastructure
4	Embrace integrated marketing communications
5	Cultivate brand partnerships
6	Balance standardisation and customisation
7	Balance global and local control
8	Establish operable guidelines
9	Implement a global brand equity measurement
10	Leverage brand elements

3.5 Local brands

A local brand is one that is unique to a relatively small number of markets. Local branding has substantial benefits

- It gets around negative country of origin images (see later in this chapter)

- It avoids confusion with other brands

- A brand name might be hard to translate in local markets or have unintended meanings

- It taps into local heritage

- If the local brand arises from an acquisition, the acquirer can maintain the brand heritage in the local brand

3.6 Global or local?

The key differences between a standardised global brand approach and an approach based upon identifying and exploiting local marketing opportunities are as described below.

(a)   Standardised global brand approach

   (i)   A standardised product offering to market segments which have exactly similar needs across cultures

   (ii)   A common approach to the marketing mix and one that is as nearly standardised as may be, given language differences

(b)   Local marketing opportunities

   (i)   A recognition that the resources of the company may be adapted to fulfil marketing opportunities in different ways, taking into account local needs and preferences but on a global basis

   (ii)   A willingness to sub-optimise the benefits of having a single global brand (eg advertising synergy) in order to optimise the benefits of meeting specific needs more closely

(a)   It is possible to move from a local brand to a global brand approach as demonstrated by the Mars Corporation with their Snickers brand. In the UK market, the biggest 'candy market' in Europe, Mars had decided to use the brand name Marathon for the chocolate bar known as Snickers in the US and elsewhere around the world. Reportedly, this was done to avoid confusion with the word knickers. There was a very distinctive brand identity in the UK to the extent that the company would sponsor the London Marathon and other sporting events to tie in with the brand name. Competition from Nestlé in Europe persuaded the company that they needed to take up the potential benefits of a standardised global brand approach rather than merely relying on a global marketing approach. They, therefore, changed the name to Snickers in the UK market at very considerable cost of advertising support.

(b)   General Motors who operate as Opel in Germany and elsewhere in Europe, and Vauxhall in the UK, chose to use the brand name Vectra for the 1996 update of the Vauxhall Cavalier in the UK market. This was done in order to standardise the name across the European market – since the 'Cavalier' model had been known as the Opel Vectra in most markets anyway. General Motors still use the Vauxhall company name in the UK market since it has strong brand equity in its own right.

For the international company marketing products which can be branded there are two further policy decisions to be made.

- The problem of deciding if and how to protect the company's brands
- Whether there should be one global brand or different national brands for a product

The major argument in favour of a single global brand is the economies of scale that it produces, both in production and promotion. But whether a global brand is the best policy (or even possible) depends on a number of factors, which address the two basic policy decisions above.

Many companies have mixture of global and local brands, marketed in a variety of ways. Branding also reflects on the image of the company not just the product. In global marketing, a company can have a variety of approaches to brands. The decision to go global or local is often a finely balanced one.

There are four choices of brand.

Individual branding for each product (solo branding)	This is the option chosen by Procter & Gamble for example, who even have different brand names within the same product line, eg Bold, Tide. The main advantage of individual product branding is that an unsuccessful brand does not adversely affect the firm's other products, nor the firm's reputation generally
Hall mark (or blanket) brand name for all product	For example, many financial services providers use the same brand for everything, for example HSBC. (There are contrasts, though. Prudential set up the Egg banking subsidiary.)
Family names for different product divisions	For example, Ford has a Premier Automotive Group, containing brands such a Saab, Jaguar and Range Rover. These are marketed individually, not as Fords.
Family/umbrella branding	For example, Kelloggs – Corn Flakes, Rice Krispies etc. This option both legitimises (because of the company name) and individualises (the individual product name).

3.7 Choice and protection of brand names

3.7.1 Legal considerations

(a)   Legal constraints may limit the possibilities for a global brand, for instance where the brand name has already been registered in a foreign country.

(b)   Protection of the brand name will often be needed, but internationally is hard to achieve.

- In some countries registration is difficult
- Brand imitation and piracy are rife in certain parts of the world

There are many examples of imitation in international branding, with products such as cigarettes, and denim jeans.

(c)   Worse still is the problem of piracy where a well known brand name is counterfeited. It is illegal in most parts of the world but in many countries there is little if any enforcement of the law. (Levis is one of the most pirated brand names.) Piracy is also a problem for intellectual property. We will return to this issue later in the chapter.

3.7.2 Considerations for brand names

The choice of brand name can at times be problematic.

(a)   **Cultural aspects of branding.** Even if a firm has no legal difficulties with branding globally, there may be cultural problems, eg unpronounceable names or names with other meanings. There are many examples of problems in global branding. But sometimes a minor spelling change is all that is needed, such as Wrigley Speermint in Germany.

(b)   **Other marketing considerations**. Many other influences affect the global branding decision, including:

- Differences between the firm's major brand and its secondary brands. The major brand is more likely to be branded globally than secondary brands.

- The importance of brand to the product sale. Where price, for example, is a more important factor, then it may not be worth the heavy expenditure needed to establish and maintain a global brand in each country; a series of national brands may be more effective.

- The problem of how to brand a product arising from acquisition or joint venture. Should the multinational company keep the name it has acquired?

Global case study

Here are examples of brand names that do not 'travel well'.

- The name 'Nova', used for GM cars, means 'it does not go' in Spanish.
- 'Ansett', the name of an Australian airline, means 'to die peacefully' in certain Chinese dialects.

Brand slogans also need to be carefully translated. Kentucky Fried Chicken found that their US slogan 'Finger licking good' could be translated in some languages to read 'Eat your fingers off'.

3.7.3 Changing brand names

Changing brand names is not a simple proposition. Most changes in brand names go from **local** to **global**. There is no point in doing so if the costs (of the changeover, depletion of brand equity in the local brand, and so on) are greater than the benefits (economies of scale, greater brand equity in the global brand, for example).

Global case study

In the UK:

- Ulay became Olay
- Munchies catfood became Brekkies
- Jif became Cif

Sometimes rebranding does not work. The consulting arm of PricewaterhouseCoopers rebranded itself as Monday. This led to adverse media comment – not helped by the fact that PwC had not garnered all the relevant domain names on the Internet, and was therefore subject to some derision. The name was abandoned.

Diageo was formed in 1997, following the merger of GrandMet and Guinness. Trading in over 180 countries around the world, the organisation required a brand name which would represent the organisation in each of these countries. At the time of their re-branding exercise there was much commentary in the marketing press about whether it was a positive or negative move. The word Diageo comes from the Latin for day (dia) and the Greek for world (geo). Diageo state that the core meaning for this name reflects the following view:

"We take this to mean every day, everywhere, people celebrate with our brands."

The name has endured so far despite the critisim at the time that it *meant nothing* to everybody.

There are four strategies to changing brand names.

Fade in/fade out	The new brand name and the old one are used simultaneously for a while, after which the old name is dropped.
Co-branding	Both brands are combined: the new global and the old local brand.
Transparent forewarning	Consumers are informed of the change: for example, HSBC, the global bank, changed all its local subsidiaries to the HSBC name and now markets itself as the world's local bank – its ads emphasise its global reach and local knowledge.
Summary axing	The old brand name is axed overnight and replaced with a new one.

3.8 The management of multinational product lines

The product assortment is often described in terms of two dimensions: width and depth.

(a)   The **width** of a product mix refers to the **number of different product lines** sold by a firm. For example, Nestlé's product lines are mainly food-related, but covers breakfast cereals, confectionery and so on. Other companies might have a more diverse mix of products.

Global case study

Vacuum cleaners

Dyson makes vacuum cleaners and washing machines. Miele, the German manufacturer, makes a wider product line, including vacuum cleaners, dishwashers, cookers and so on.

(b)   The **depth** of the product line refers to the number of items sold within a product line. For example, Nestlé may sell many different types of breakfast cereals, in different packages.

The factors that determine **width and depth** are as follows.

Size and focus of the company	A small company might only sell a few products. A large company, such as Nokia, may still focus on a narrow product line but produce many variants. Large conglomerates sell a variety of products and services.
Nature of the business	Banks sell bank accounts, savings accounts, loans and, in some countries, insurance products. There may be restrictions as to what they can actually sell.
	Airlines sell air transportation, and their main economic task is to manage capacity for a product. Some sell different types of seat to different customers, and even organise holidays. Others, such as easyJet, pride themselves on selling a no-frills service. A deep product line for an airline would be a large number of different destinations and seat types.
	Many publishers sell a wide variety of books for leisure and work. They brand them separately (by developing imprints).
	Virgin is branded over cola, rail transport, air travel, music sales and mobile communications

Customer preferences	Customers in some countries may have preferences that others do not share so the product line will be adjusted accordingly
Price spectrum	Premium and budget products might be offered, or different sizes of pack at different levels
Organisational structure	With MNCs that are organised on a country basis, product lines may evolve independently in the different countries.
Profitability	Accounting techniques such as direct product profitability can influence choice whether to carry on with or drop a product. Such decisions need to be made carefully.
Competitive environment	Where competitors are already well entrenched, it may be very difficult for newcomers to offer comprehensive product lines.

3.9 International piracy and breach of copyright

3.9.1 Diluting the brand: legal threats to uniqueness

A brand is a **unique** identity offering certain 'promises' about the product. Brand owners spend a great deal of money investing in their brands. If they are popular, or can justify premium prices, then **other suppliers** might wish to offer a similar benefit. Some are quite unscrupulous in how they do so. There are several different kinds of appropriation.

Global case study

(1)   Microsoft vs Apple

In the early 1990s, when Windows was being developed with a graphic user interface, Apple sued Microsoft, claiming that Windows copied Apple's 'look and feel'. Apple was unsuccessful.

(2)   Own label brands

Own-label brands (also known as private-label brands) are becoming increasingly popular. Tesco, the UK's largest retailer, offers a **value** range, with simple packaging, a **standard** range and a **premium** range (Tesco's Finest). However, this brand building has taken some time, and it is not clear that private-label brands exert quite the extent of power that they do in the UK.

The retailers obviously keep to legal guidelines. Even so, in some rare occasions, manufacturers have been concerned that the own-label packaging is too similar to the branded product.

It is possible to acquire fakes of watches, clothing and so on in some parts of China. Most purchasers know they are fakes, but are happy with the imitation.

Activity 3

Many manufacturers of brands also manufacture for own-label producers (private-label). Do you think this policy suggests that branding is less important to manufacturers than they claim?

Where products, and the benefits they offer, **can be easily (and legally) imitated**, then manufacturers need to consider other strategies to deal with this.

  Developing Integrated Communications Strategy

Brand building	There may be little functional difference between **own label** and **branded products**, but different areas of the marketing mix can be used to support the branded products (eg advertising).
Innovation	New or improved versions of products, or with new packaging, can be offered to keep a step ahead of the competition.
Process innovation	A **product** might be easily copied, but processes can be harder to copy. A firm with a unique service process may do better than others offering similar benefits.

3.10 International brand piracy

While there are grey areas, as identified above, there have been real problems with **illegal copying** and **piracy**.

The brand name	A fake product can be sold under the brand name. For example, fake Rolex watches are easy to buy in some countries.
Logo	A logo can be similar to another.
Design	This is little harder. Some design elements might be covered by patents (which offer life time protection). In other cases a product might be 'reverse engineered'
Packaging	Similar or identical shapes can be used in packaging (for example, bottles).
The product itself	Pirate versions of films on DVD, software; illegal file sharing of music downloads.

3.10.1 Implications of piracy

The implications of piracy are that a consumer might be misled and might believe that the product he or she has purchased is the 'real thing'.

Financial losses	A firm might lose overall sales if people buy the pirated version. If the pirated version is of sufficient quality, then the consumer gets the benefit for less money. Pirated CDs and DVDs may be 'good enough'. Pirated software may be full of bugs, however.
Brand name	Generally, pirated products are of poorer quality. Where people **do not know** they have bought pirated products, then the reputation of the brand owner will suffer. Examples are counterfeit medicines or pirated car parts. If these are of poor quality, then public safety is at risk.

It has become easier to pirate products than before.

- The spread of advanced manufacturing technology means that passable imitations are easier to produce.

- Intellectual property in digital form does not lose quality when it is copied from one medium to another.

- There are high profit margins in pirated products.

In some cases, unauthorised distributors can infiltrate copies into the supply chain.

Activity 4

We mentioned the example of fake watches. In some parts of the world, these are sold at very low prices indeed. How is Rolex harmed by low-price imitations?

3.10.2 Dealing with piracy

Litigation	Legal action can be taken. Some record companies have sued teenagers for making illegal downloads: this is partly to discourage others from doing the same. This can be a double-edged strategy, if it generates sympathy. Where local government and law enforcement institutions connive in the process, it harder to take action.
Lobbying	Firms can petition their government to put pressure on other governments to take stronger enforcement action.
Customs and excise	Customs officers sometimes take samples of product to check for compliance.
Product	It might be possible to introduce unique identifiers to the product. An alternative is to make a low price 'value' range
Distributors	Changing the distribution strategy might help

3.11 'Country of origin' (COO)

> **Definition**
>
> **Country of origin.** For many products, the country in which a product was produced can evoke certain associations, either positive or negative. These perceptions can act as a cue to evaluate products. In other words, consumers' cultural stereotypes of countries influence their assessment of products and subsequent purchase.

Branding, marketing communications and product adoption by consumers can be influenced by the country of origin of the product that is being sold to them. 'Made in Germany' for example, may symbolise quality. Country of origin influences may affect the nature of the product or the nature of the marketing messages.

Country of origin is important to international marketing for a variety of reasons.

Legal: as a guarantee of authenticity	The EU has strict regimes for geographical identification. Feta cheese comes from Greece, and a Danish firm making 'feta cheese' now cannot market it as such.
National stereotypes, eg as an indicator of quality or other values	A country may be synonymous with a quality products. For example, German cars are often held to have a reputation of high quality. Smirnoff vodka is branded as if it were a Russian brand however it is actually a British brand. Along similar lines, consumers would expect Bertoli pasta sauces to be made in Italy when indeed they are manufactured in Holland.
Patriotism: domestic v foreign products	People might buy certain products for patriotic reasons – or rather certain market segments might invoke patriotism in their buying behaviour. For example, 'French Fries' were renamed 'Freedom Fries' in the US, after France's opposition to US policy on Iraq. Patriotism as a buying motive should not be overstated.
Essential to the product or service	Tourism is an industry where 'country of origin' effects are vital. Tourist agencies brand their countries to appeal to different market segments. New Zealand was the location for the epic film Lord of the Rings and could, perhaps, market itself as 'Middle Earth'. Malaysia has had a long running campaign branding itself as Malaysia Truly Asia.

Essential to the brand	IKEA is a global company, but the vast majority of its products are branded with Swedish names. The colour coding (blue and yellow) is the colour of the Swedish flag. The products are probably packaged all over the world (and assembled at home). Stores across the world only vary according to different legal trading regulations of different countries. Interestingly however, the consumer perception of IKEA can vary considerably between countries. During the 1990's when IKEA began to export the brand to the UK, Swedish consumers regarded products as cheap and possibly inferior whilst UK consumers regarded it as innovative, good value and higher quality compared to alternatives within the market.
A signalling cue	It can help a company make its offering distinctive. Country of origin effects are exploited in alcoholic drinks (the Australian origin of two beers, Foster and Castlemain XXXX, even though these are brewed under licence).
A problem to be dealt with	**To be hidden?** Some companies like to obscure their national origin or ownership as to do so makes marketing or operations easier. If political relations are uncertain, then a company might hide behind a local identity. This can mean significant adaptation. **To be overcome?** In other cases, national stereotypes have to be overcome (eg that products from certain countries are expensive) through advertising or the use of local partners. **To be diluted?** Increase the share of a product made locally. This can be important in winning government contracts (eg defence, that there is some technology transfer). **To be taken out of the equation?** Build up global brands where national origin is ignored. Vodafone, Orange, T-mobile and other phones do not have strong 'national' images as they are global services. They operate global partnerships, however. The same is not true of airlines (British Airlines, Air France etc) where nationality comes into play. BA tried to reduce its perceived Britishness by redesigning the tailfins on its aircraft. This generated quite a hostile response, including criticism from the Prime Minister of the day. When discussing a possible merger, BA and Iberia were keen to announce that they were to retain their own individual branding.

French wine

Wine is a growth industry in Australia, Chile and many East European countries, but it is suffering in France (as you saw in the earlier Global Case study). France is the home of wine but local consumption is falling and where overseas market share is declining. There are some simple product reasons for this – some consumers buy by grape variety not the complicated French terroir system, and 'brand' reliability is important. For example, people know what to expect from Jacobs Creek, an Australian wine – this is a far cry from experts who are able to tell which is a good year for 'Bordeaux'. France's system of appellation controllée makes it harder to market to consumers in north Europe.

3.11.1 The variety of COO effects

The table above referred to individual products. In practice, though, branding and communications practices might be influenced by a number of COO effects. Some may be deliberate policy; other effects may be unintentional.

Usunier and Lee (2005) suggest five different factors underpinning country of origin, that reflect on the perceptions of the 'nationality' of the product.

1 Global image of the product in terms of domestic goods and foreign goods

Imported versus home produced products: people might assume that all imports are better (or worse)	In general people might have **generic** views of the relative benefits of **imported** as opposed to home-produced products. Where domestic product quality is low, any import may be a badge of quality. (The converse may be true: people may not trust imports. For example, they may assume that any imported products are of poorer quality than that produced at home.) Consider services, too. It is possible to pay for medical treatment quite cheaply in certain countries (eg Cuba), where 'medical tourism' can be encouraged. Some British people may be happy to visit a Hungarian dentist – on the grounds of cheapness. There is an implicit trust in equivalent quality.
'National' versus 'international' products	A generic product may be associated with a national image (not necessarily that of the producer!) or not associated with a national image. For example, camera film is not a 'national' product. Cultural products, such as movies, might have national connotations. 'Chinese medicine' in the West is promoted as an alternative to western medicine. Ayurvedic medicine is promoted on the basis of its Indian origins. As these are specific treatments and procedures, they are 'national' rather than international products.

2 Country image of the generic product

Does a product 'signify' a particular country? (For example, do you consider that the Harry Potter movies trade on their Englishness?) This is relatively rare.

Neutral	Some products have no country image at all. Credit cards are a generic product and are not associated with a particular country. (The name Visa suggests an international outlook. Contrast with American Express, which has a distinctive branding.)
A specific national image	Scotch whisky trades heavily on its unique Scottish heritage, even though whiskies are made elsewhere, in Japan for example.
Connotes several different countries	Tea is connoted by India, Sri Lanka, China and some African countries.
Regional image	Wine is not a country-specific product, but it is does have geographical associations (eg 'old world' versus 'new world' wines).

3 The national image of the producer

Image related to nationality	L'Oréal perfume is obviously connected to France, as it has a French name.
Corporate image	Most firms have some sort of corporate identity. Some airlines often trade on national origin in their corporate image.

4 Country evoked by the brand name

Neutral	The name Kodak is supposed to have no linguistic or cultural connotations. (The same cannot be said of Fuji, its main competitor, although Fuji does not use its Japanese heritage in its advertising.) Other names can evoke countries (eg Mitsubishi).
Specific linguistic/country connotations	Every IKEA store is decked out in the blue and yellow of the Swedish flag. All the products are given Swedish names. IKEA offers 'a taste of Sweden' and offers Swedish foods.

5 Place of manufacture – 'made in' label

Generally, the 'made in' label shows where a product was actually produced. However, the product may have been made from imported components and assembled on site. Therefore, the label may tell us less than we might think.

Global case study

These COO elements are combined in a number of different ways. Usunier and Lee (2005, p. 287) cite the example of Kinder – milk chocolate bars made by the Italian firm Ferrero (who also make Belgian chocolates under the Ferrero Rocher brand). Chocolate as a product evokes 'Switzerland', apparently. Ferrero's name evokes Italy. The name of the chocolate bar, Kinder, is German for children. The 'made in' label is hardly visible.

3.11.2 Practical uses

Usunier and Lee (2005) propose four strategies for dealing with country of origin effects.

National image of the generic product	A firm should 'diffuse an image which corresponds in each country to what is locally valued (imported or national) in the product category concerned'. In other words, the producer should look at existing perceptions in the target market.
National image of the manufacturer	If the product category is generally associated with a specific country of origin, one should change the brand name accordingly. For example, a manufacturer of machine tools should not be reluctant to adopt a German name, because of the favourable association of German-sounding names with technical reliability.

Country in the brand name	Countries represent things to other consumers. The visibility of the company name, the brand name and the 'made in ...' label should be adjusted depending on their respective ability to convey the desired symbolic meanings.
Country in the 'made in ...' label	Where negative connotations exist, the advice is not draw attention to the actual country of manufacture. Negative COO perceptions would suggest reducing the size of the label. More positive connotations would imply enlarging the size of the label.

3.11.3 Consumer product evaluation

Country of origin is salient if it has an **association** (eg Germany = quality, France = hedonism). COO factors were often considered in isolation, whereas in fact COO is one of several factors in a customer's experience of a product.

COO has a stronger effect at the **earlier stages of the buying process**, underpinning views as to brand quality as opposed to purchase intention. COO is also used as a cue or shorthand for the quality of the brand.

Activity 5

Which?, the magazine of the Consumers Association – a UK consumer pressure group – reported a decline in the perceived quality of German cars such as Mercedes. Do you think there may be implications for other German consumer brands sold abroad?

3.11.4 COO and national stereotypes

Usunier and Lee (2005) suggest that national stereotypes vary. Consumers may be unwilling to buy from countries they 'do not like' for a variety of reasons, for example government policy. They also note that COO stereotypes are less powerful once the importer gets used to the local supplier. In other words, a learning process reduces the impact. Country images change over time. The best example is that of Japan, since WWII, and, more recently, of South Korea in overcoming negative images of quality.

However, the influence of COO effects can be reduced for a number of reasons.

Customer awareness of COO	Global brands, such as Sony, are manufactured all over the world: the brand is powerful enough. In other cases, 'made in' has a bigger impact than the brand name. Country of design, rather than country of manufacture, may be an important concern for some products (as many products subcomponents are subcontracted).
	For example, Dyson vacuum cleaners are designed in the UK, but manufactured in Malaysia.
Customer knowledge of the product category	Expert consumers base their evaluation on the strength of actual product attributes, whereas novices tend to rely on the COO cue as such. The halo effect is used where consumers are unfamiliar with the product category.
Consumers' ethnocentric tendencies	We note in the case study on mountain bikes below that COO effects vary on the individualist/collectivist dimension of Hofstede's classification.

The product category	COO effects do not vary between consumer and industrial goods. However, COO effects may be limited to product categories (eg people might associate French perfume positively but French wine negatively).
Perceptions of risk	National products may seem less risky – for example if this includes after sales service and support.

Malaysia

Mohamed et al (2000) describe the susceptibility of Malaysian consumers to COO preferences, in relation to the product. Attitudes towards locally made goods were determined by price, style and availability. More expensive 'designer' products are influenced by COO.

Malaysians preferred goods originating the USA, Japan, France and Italy to home-manufactured products, but preferred Malaysian products to others produced in South East Asia.

Although it is possible to speak in general terms about possible country of origin effects, is there any evidence, other than anecdotal, that these do actually apply? Are some cultures more than others particularly susceptible to country of origin effects?

Mountain bikes

COO effects and the Hofstede classification

Gurhan-Canli and Maheswaran (2000) argue that the strength of COO effects is not universal and varies from country to country. COO effects are stronger than single 'animosity' or affection towards a particular culture or country. We can use Hofstede's individualism-collectivism continuum to refine the discussion. A culture can be differentiated on the basis of whether it deals with horizontal or vertical relationships. In 'vertical' culture, there may be an emphasis on hierarchy. In a horizontal culture, membership of a group is a matter of **equality**: there are few distinctions based on official status, and perhaps communication pathways are networked.

They took the example of mountain bikes and compared these with different consumers.

The authors argue that COO effects:

- Vary across cultures
- Can be modelled on the individualism-collectivism continuum of the Hofstede classification
- Reflect the respect for hierarchy in different cultures

They compare the USA with Japan and, after a variety of tests came to the following conclusion.

Individualist 'cultures' prefer the 'home' country product only when it is technically superior to the competition. 'Collectivists', where there might be a greater orientation towards the group, tend to prefer the home product even though, on a crude comparison of product attributes, it is inferior.

Implications for marketers

Domestic producers in some countries can benefit from advertising COO to their own markets. However, it may not be worthwhile in other markets. COO will only matter in 'individualist' cultures if the product is, in fact, superior: US consumers choose foreign products on the basis of perceived quality.

3.11.5 COO and global organisations

COO effects become more complicated with global integration. For example, Sony products can be made in a number of countries from different parts. Some countries try to avoid 'made in' labels entirely.

Brandscape Africa has some very interesting resources relating to the branding of Africa and African products. The resources available refer to the perception of African brands both within Africa and outside in the wider world. Within the top brands recognised within Africa (the top five according to Superbrands East Africa being Nokia, Blue Band, OMO, Coca-Cola and Colgate), none are actually produced in Africa. Why this is occurring and strategies to strengthen branding by large and SME African companies are discussed. Go to:
http://www.brandscapeafrica.org/index.php/resources.html

3.12 Strategies for dealing with COO effects

Mohamed et al (2000) provide a useful table based on the 4 Ps of the marketing mix.

	Positive COO effects	Negative COO effects
Product	Emphasise 'made in'	Emphasise brand, not country
Price	Premium price, to benefit from COO effect	Low price, to compete those for whom value is important
Place	Exclusive distribution	Partnership in the supply chain, to downgrade the importance of locations
Promotion	Country image in promotions to reinforce COO advantage	Brand image should be developed

Global case study

Malaysia is the world's leader in the manufacture of latex medical gloves. The Standard Malaysian Glove programme is a joint effort between the Malaysian Rubber Board and the Malaysian Rubber Glove Manufacturers' Association. The programme was developed in consultation with regulatory and standard setting bodies such as the US FDA.

The programme is designed to set standards for barrier performance, protein (linked to allergic reactions to latex gloves) and powder (containing protein) levels.

The SMG logo is used as a sign of quality assurance. The green version of the logo denotes powder free variants and the orange logo is used for a lightly powdered variant.

The relevance of the logo is that it acts as a short cut symbol to reassure users that gloves are superior in barrier protection with the reduced risk of protein sensitisation. Manufacturers must be accredited by the Malaysian Rubber Board and all must have an equivalent quality management system in place such as ISO 9001-2000.

Based on: http://www.mrepc.com/smg/

1 Discuss how to manage brands over time

- Over the course of the product life cycle brands may need refreshing, or extending in order to lengthen the lifespan.

- There is need to maintain core brand values over time to build strong brands.

2 Suggest alternative methods of measuring and valuing brands

- Brand valuation has become more widely used since it has been possible to quantify intangible assets in financial reporting.

- A number of research methods can be used to collect information – both qualitative and quantitative measures are applicable.

- A large number of proprietor methodologies exist to value brands.

3 Identify the complexities of global branding decisions

- A **product** is the satisfaction purchased by the buyer. The same physical object, however, can have different uses and meanings.

- International marketing, the choice is between **standardising** the product in all markets to reap the advantage of scale economies in manufacture and, on the other hand, **adaptation** gives the advantages of flexible response to local market conditions. Adaptation may be unavoidable in some circumstances (eg if different countries have different legal standards for safety).

- Packaging, labelling – and even branding – must sometimes be sensitive to **local conditions**.

- A product's stage on the **product life cycle** may vary from country to country, and different marketing strategies will be appropriate in each case.

- **Time based competition** is becoming increasingly important in **new product development**. Product development may be co-ordinated across several international markets.

- Different product types may require different **launch strategies**.

- Companies may have a single global brand, because of the economies of scale that may be realised, but **global branding** may not be the best strategy.

1   This will depend on your own research

2   This will depend on your own organisation. Please remember however that there are many agencies who specialise in valuing brands – as yet there is not one best approach.

3   Some manufacturers such as Kelloggs outwardly state that they do not produce retailer own brands so that the image of the brand is not cheapened. The products formulated for private labels do not normally have the same formulations and therefore the brand retains their unique qualities.

4   Rolex suffers when unwitting consumers believe that they really have purchased the real product (and the quality is exceptionally poor).

5   Depending on whether consumers perceived the issue to be specifically Mercedes related or not may determine the effect on other German products. Mercedes brand values are closely aligned to German engineering quality and so if there was no corrective action taken by Mercedes to improve quality then there is a possibility of an adverse effect in the longer term.

Arnold, D., (1992). *The Handbook of Brand Management*. Reading, MA: Addison-Wesley.

Douglas, S.P., (2001). Executive Insights: Integrating Branding Strategies across Markets: Building International Brand Architecture, *Journal of International Marketing*, Vol. 9, No 2 pp. 97-114.

Gurhan–Canli, Z. and Maheswaran, D., (2000). 'Cultural variations in country of origin effects', *Journal of Marketing Research*, Vol. 37, Issue 3 pp. 309-317.

Keller, K., (2008). *Strategic Brand Management.* International edition. 3rd edition. Harlow: Prentice Hall.

Kotabe, M. and Helsen, K., (2004) *Global Marketing Management*, Hoboken NJ., John Wiley and Sons Inc.

Leonidou, L., (1996) Product Standardization or Adaptation: The Japanese Approach *Journal of Marketing Practice: Applied Marketing Science*, Vol 2. No 4 pp. 53-77.

Mohamed, O. et al, (2000) Does 'made in...' matter to consumers? A Malaysian study of country of origin effect. *Multinational Business Review*, Fall 2000. Vol. 8, Issue 2, pp. 69-73.

Usunier, J.C . and Lee, J., (2005) *Marketing Across Cultures.* Harlow: Financial Times, Prentice Hall.

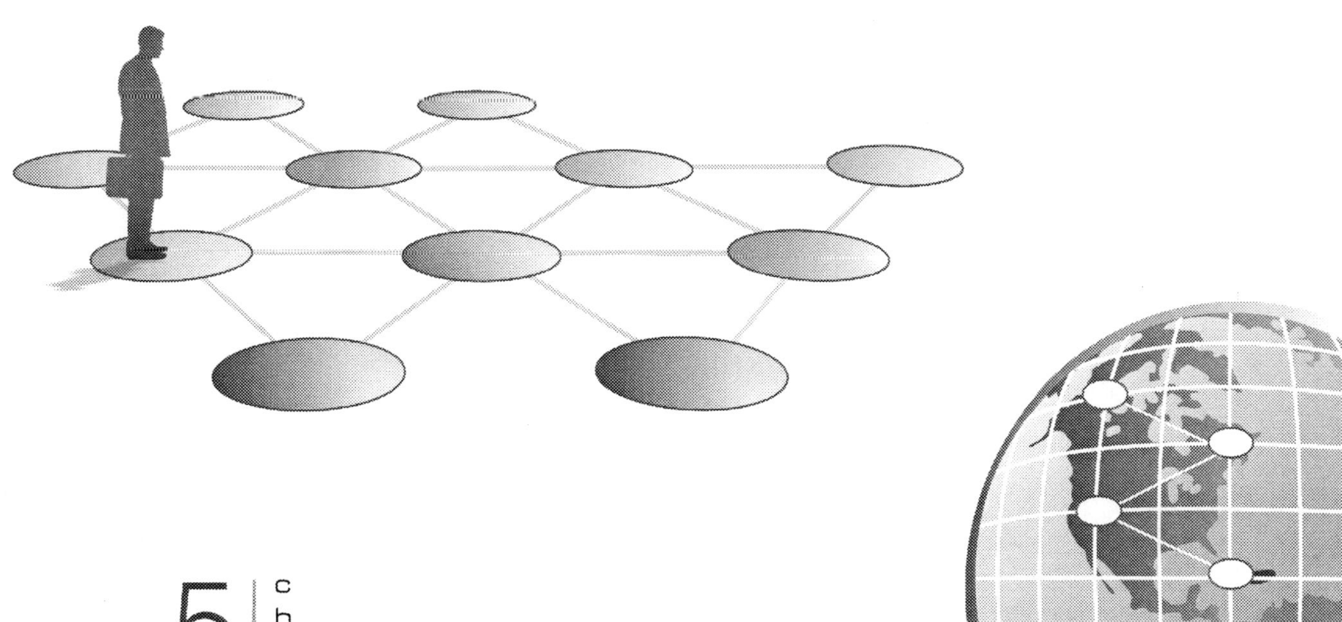

5 | chapter

The relevance of digital marketing

Within this short chapter we begin our discussions about the importance and relevance of digital marketing. Unfortunately the nature of the digital marketplace means that we are almost certain that by the time this text is printed, some of the information will already be out of date! The almost frantic rate of change within the technological world is not likely to slow anytime soon. The important aspect to remember however is that good marketing practices should outlive the changes in technology in the respect that good marketing should mean meeting customers needs. Digital marketing offers plenty of scope for finding new and often improved ways of doing this but the key definition of 'marketing' should remain true. With that in mind digital marketing can be regarded simply as an alternative (albeit sophisticated) medium at marketers' disposal.

The chapter looks initially at how digital marketing can be integrated into the wider marketing plan. It then moves to consider the functions of the e-marketplace and the e-marketing mix.

This chapter is deliberately theoretical in nature because it precedes Chapter 6 which is written by an expert digital practitioner. Chapter 7 briefly summarises any remaining theoretical points concerning e-marketing.

Contents

By the end of this theory based chapter you should be able to:

- Discuss how digital marketing integrates with the wider marketing plan

- Describe the e-marketplace and discuss things to consider with regard to online consumers

- Evaluate the benefits of digital marketing

1 Digital marketing within the marketing plan

By this point in the study text, it should be no surprise to you that the digital marketing employed should be well integrated with the rest of the marketing plan. Digital marketing is sometimes referred to as Web 2.0. The central premise of integrated marketing requires that all marketing is consistent and all elements work together to better meet customer needs. Over recent years there has been an implicit expectation that marketers employ digital marketing techniques. Sometimes this has been to the detriment of the marketing message because digital campaigns have been designed in isolation.

Global case study

A 2009 survey by US recruitment agency Heidrick and Struggles with 111 senior marketers working within firms with annual revenues over $1billion showed that marketers see digital as the medium of choice but that they are not achieving the results they require from it.

Only 16% of respondents were very satisfied with their execution of digital campaigns in terms of their ability to respond quickly to new opportunities.

Lynne Seid, partner in Heidrick and Struggles' global consumer practice said the big takeaway from the survey is that there's still enormous room for improvement for most companies' digital marketing strategies.

"What I'm hearing anecdotally is there are now sometimes half a dozen digital agencies and suppliers specializing in social media and search," Seid said. "We don't have anyone managing, integrating and demanding best practices in those areas." Seid envisions a "digital CMO" taking responsibility for managing those disciplines. Said Seid: "That will be the CMO of the future."

Based on: Brand Week (2009).

http://www.brandweek.com/bw/content_display/news-and-features/digital/e3if46ca983d59bcb8f9823719f0e98475c?pn=2

The development of digital marketing tools is not likely to slow at any point soon. Although over the last few years the focus for some marketers has been driven by 'availability' of the tools and need for 'experimentation' rather than by a disciplined approach to reaching the target market through an integrated and consistent approach. Sometimes the simplest marketing ideas are the most effective, and these are not necessarily the most technologically advanced.

> **Definition**
>
> **Digital marketing** has been defined by Jobber (2007, p.723) as follows:
>
> *"the application of digital technologies that form channels to market (the Internet, mobile communications, interactive television and wireless) to achieve corporate goals through meeting and exceeding customer needs better than the competition. ICT is adding impact, speed, interactivity and fun to the full range of promotional methods and tools."*

Digital technologies and the imaginative use of **websites** are creating endless possibilities. Since the first live website in 1991: http://info.cern.ch, over a billion people throughout the world now regularly use the web to find products, entertainment, soul mates and conduct business (Chaffey et al 2009).

The Internet has transformed marketing from a number of perspectives and has opened up vast opportunities for both businesses and consumers. For business opportunities include:

- The promotional opportunities which now exist.
- The ability to offer new services at relatively low cost
- Application of online communication techniques
- Opening of competition so smaller organisations are able to compete within the market

The changes for consumers has meant:

- increased product and service choice with direct global access to online sellers

- involvement in more aspects of service delivery eg online check in for airlines

- opportunities for easier research during the information seeking stage when making purchase decisions eg accessible information, access to reviews from existing consumers, online discussion forums to gather third party opinions.

1.1.1 The promotional mix perspective

A number of new promotional opportunities are now available or have changed as a result of the Internet.

(a)   **Advertising**: using direct response advertising, web-advertising, CD-ROM and video packages, podcasts, viral campaigns and mobile phone advertisements, ranking on Internet search engines

(b)   **Direct marketing**: using email or mobile text messages instead of conventional mail shots

(c)   **Sales promotion**: online vouchers, discounts, loyalty schemes, 'SMS to win', competitions

(d)   **Public and media relations**: corporate image on websites, posting of online press releases, special areas of the website for trade/press/client, news bulletins, customer review websites, crisis management, publicity for exhibitions and events

(e)   **Point of sale display** at online shopping sites

(f)   **Personal selling**: connecting mobile sales forces to customer/product databases and sales tools (eg video or computer modelling on the sales person's laptop, demonstrating product use or performance)

(g)   **Relationship marketing**: generating multiple contacts via website, email, phone 'remembering' customer details and preferences; allowing customer service staff to 'recognise' callers with relevant data. Social media has however offered a opportunity to identity and help develop relationships

(h)   **Customer loyalty programmes**: value-added benefits that enhance the Internet buying experience, user home page customisation, virtual communities (chat rooms etc), SMS messaging

Guerrilla marketing was first used as a term by Jay Conrad Levinson in 1984. He describes guerrilla marketing:

"I'm referring to the soul and essence of guerrilla marketing which remain as always – achieving conventional goals, such as profits and joy, with unconventional methods, such as investing energy instead of money." Levinson (2009).

Campaigns have included activities such as:

- Virgin airlines placing a tray of fresh eggs on an airport baggage collection conveyer belt to demonstrate their care and attention to safety.

- VW placing an ice sculpture of their Polo model outside the Saatchi Gallery in London in 2004 as part of the origination for artwork and launch of the four-week national press advertising campaign for the Twist. The model was branded as 'chilled' with free air conditioning – hence the ice sculpture which attracted significant media coverage in its own right.

- McDonalds using a giant egg billboard in Chicago which cracked open each morning demonstrating that McDonalds use fresh eggs in their breakfasts.

Some key campaigns of 2008 are outlined in the following marketing blog:

http://daveibsen.typepad.com/5_blogs_before_lunch/2008/12/best-guerrilla-marketing-campaigns-of-2008.html

Guerilla marketing became hugely popular in the first decade of the 2000's and continue to be used imaginatively. Increasingly however many guerrilla markting campaigns appear to have been replaced with cheaper and more widely availible viral marketing. The infamous Cadbury drumming gorilla (ape) and their children with dancing eyebrows were possibly more effective than any offline guerilla campaigns because of their ability to reach vast audiences at incredible speed. If you were one of the few people on the planet to miss these campaigns they can be found at:

http://www.aglassandahalffullproductions.com

http://www.youtube.com/watch?v=TnzFRV1Lwlo

http://www.youtube.com/watch?v=TVblWq3tDwY&feature=related

1.2 Common digital tools

We will return to specific digital tools in Chapters 6 and 7 however it is useful to mention typical digital tools used so that you are aware of the range of key digital media tools.

Digital tool	Brief explanation
Corporate or brand website	Types of online presence includes: 1. **Transactional e-commerce site** – products are purchased online eg www.amazon.com 2. **Services-orientated relationship building website** – information to stimulate purchases and assist decision-making eg www.pwcglobal.com 3. **Brand-building site** – an experience to support the brand. Not for purchases generally. eg www.cadbury.co.uk, www.guinness.com 4. **Portal or media site** – intermediary sites providing news, search engine, directories, shopping comparisons eg www.google.com, www.FT.com

	5. **Social network or community site** – enabling interactions between different consumers, often these sites are funded by advertising revenues earned eg www.Facebook.com, www.myspace.com, www.youtube.com
Search engine marketing (SEM)	Search engine optimisation boosts a company's position in search listing. For example you may use the search term 'digital marketing agency' in Google, Yahoo or another search engine. The agency website listed at the top has the highest chance of being clicked on. Paid search marketing uses sponsored ads which are paid for on a Pay Per Click basis.
Online PR	Maximising favourable mentions of your brands, products, company or website on a range of third party sites, social networks and blogs. Can help to raise brand awareness.
Online partnerships	Long term arrangements to promote via a third party website or digital media resource through co-branding, affiliate marketing, link building.
Interactive advertising	Online ads including banner ads, skyscrapers, rich media ads to encourage 'click throughs'
Opt-in email	Renting email lists – placing ads in third party newsletters, building an in-house email list, sending e-newsletters and email campaigns
Viral marketing	Online word of mouth – messages are forwarded by consumers – messages tend to have a link to a site with jokes, information, pictures and videos

1.3 The Internet and the marketing mix

Brassington & Pettitt (2003) itemise the marketing uses of the Internet. They distinguish between the Internet for assistance in research and planning, distribution and customer service and communication and promotion tools. We have added a few more points to their original list:

As a research and planning tool

- Obtain market information
- Conduct primary research
- Analyse customer feedback and responses
- Valuable secondary sources of information

Distribution and customer service

- Take orders
- Update product offerings
- Help the customer to buy
- Process payments
- Raise customer service levels
- Reduce marketing and distribution costs
- Distribute digital products (music, software etc)
- Encourage 'prosumers' – consumers who are actively involved in the design of products

Communication and promotion

- Generate enquiries
- Enable low-cost direct communication
- Reinforce corporate identity and present company in a good light
- Produce and display product catalogues and product information
- Entertain, amuse and build goodwill
- Inform investors, suppliers and employees of developments

Using Brassington and Pettitt's (2003) list of Internet marketing uses, conduct a quick audit of practices within your own organisation.

In your opinion are there any areas which could be improved?

Jobber (2007) talks about how the evolving digital business environment has facilitated 'many-to-many' communications. Chapter 18 of his book contains accessible content on marketing in the digital age.

Assessment advice

You should of course be aware of these technological developments, but also realistic about their current application. It is important for marketers to be aware of what ICT – especially the Internet – can't do. It may be able to deliver some products/services in 'real time', or very fast: information, music and images, educational material and banking transactions.

However, many products will still have to be physically delivered. The Internet is global in its reach, and so products will have to be delivered internationally. This takes resources, logistics, infrastructure and time.

Because of the promotional strengths of the Internet, there is great potential for customer disappointment if the product does not live up to the sophistication of the promises – or if it cannot be delivered in a reasonable condition or within a reasonable period of time.

1.4 Linking digital marketing with the overall marketing plan

There are arguments both for an against the need for a separate digital marketing plan. Chaffey et al (2009) argue that in many organisations a distinct e-marketing plan is initially essential if the organisation is to effectively harness digital marketing. However, the authors continue to state that once an organisation has successfully defined its approaches to internet marketing, it is likely that the Internet can be considered as any other communication medium and included within the overall marketing plan.

The following problems are typical when there is no clear plan for e-marketing:

- Customer **demand** for online services is unknown

- **Competitors will gain market share** if insufficient resources are devoted to e-marketing and clear strategies are not defined

- **Resources will be duplicated** with similar activities planned for online and offline marketing but there may also be a lack of e-specialist skills

- Customer **data collected online** may not be well integrated with other systems used to manage customer relationships

- **Efficiencies** available through online marketing will be missed

- E-marketing will not be viewed as a major strategic initiative if plans do not filter to **senior managers.**

The key point with a separate plan according to Chaffey et al (2009) is that it can in many instances be just a two page summary or annex to the overall marketing plan but it must:

"set clear objectives and strategies showing how the digital presence should contribute to the sales and marketing process. Specific initiatives that are required such as search marketing, email marketing or features of website redesign can be specified" p. 210

Dave Chaffey, the e-marketing expert and joint author of two of the module recommended texts, has an active website with a number of useful resources. The following article refers to key questions that should be asked when designing e-strategy:

http://www.davechaffey.com/E-marketing-Insights/Internet-marketing-articles/Top-10-E-marketing-strategies

1.5 SOSTAC® as a generic strategic approach

Regardless of whether an organisation has separate or combined online and offline marketing plans, they need to be carefully researched with fully justified suggested actions with clear guidance for the implementation of activities and metrics to ensure results are measured against the original objectives. SOSTAC® is an approach used widely within strategic marketing planning. Designed by PR Smith (1993), Chaffey and Smith (2008) recommend using the SOSTAC® approach to plan e-marketing strategy. Many marketing texts explain how to use the SOSTAC® method and advocate its use.

Activity 2

How are marketing plans constructed within your own organisation? Are any frameworks such as SOSTAC® employed?

PR Smith is a co-author of Chaffey and Smith (1998) and Smith & Taylor (2007) – two of the module recommended reading texts. PR Smith designed the SOSTAC® method of marketing planning. He consults widely on how to integrate the model into organisations planning. His website looks briefly at the SOSTAC® model but has a Wall of Fame with video clips of several leading marketing academics and practitioners including Philip Kotler on branding. View the website at:

http://www.prsmith.org

When we look at integrated marketing planning in later chapters, we will refer back to SOSTAC® from a more generic perspective.

1.5.1 The SOSTAC® acronym

SOSTAC® refers to the elements required within a marketing (or e-marketing) plan. This is highly consistent with any planning, and you should be familiar with these main concepts from earlier fundamental marketing courses or reading.

Definition

SOSTAC®

Situation analysis, objectives, strategy, tactics, actions and control.

The table below highlights the implications for digital marketing.

Initial	Refers to:	Meaning	e-implications
S	Situation analysis	**Where are we now?** A complete audit is required. Remember PESTEL and SWOT analysis Internal analysis Competitor analysis Brand audit Customer insight	How many customers are buying or influenced online? What is the forecast growth? What is changing in the online world? Focus on the e-market place with a specific e-market SWOT along with an audit of current e-capabilities.
O	Objectives	**Where do we want to be?** Should be specific, measurable, actionable, realistic and within a timescale.	Why go online? What are the benefits and what is the purpose of going to the effort? **5Ss** – Chaffey and Smith's (2008) Objectives of digital marketing: **Sell** – customer acquisition and retention targets **Serve** – customer satisfaction targets **Sizzle** – site stickiness – length of visit **Speak** – number of engaged customers **Save** – quantified efficiency gains
S	Strategy	**How do we get there?** Summarises how to fulfil objectives.	Which trends are we responding to? Which media mix, which targets? Digital marketing mix, type of online site? Detail: Segmentation, target and positioning Online value proposition Integration with other communications Tools (web, email, interactive TV, intranet, Internet etc)
T	Tactics	**How exactly do we get there?** Tactics explain how to implement the strategy.	Break the tactical e-tools into more detail, what level of integration between tools. E-marketing mix (communications, social networking, interactive TV, web design and functionality)
A	Actions	**What are the details of the tactics?** Each tactic should be treated and planned as a mini project. Internal marketing should be used to explain and motivate staff to execute the actions. Who does what and when?	eg what actions have to be taken to create a banner ad, interactive TV ad, opt in email campaign. Detail: Responsibilities and structures Internal resources and skills requirements External agencies

 Developing Integrated Communications Strategy

Initial	Refers to:	Meaning	e-implications
C	Control	**How do we monitor performance?** What metrics are used and how are they used to measure success or failure? What actions (eg abort project) are to be taken in which circumstances?	e-based Key Performance Indicators and metrics: Usability of website – mystery shopping Site visitor profiling Satisfaction surveys Frequency and process for reporting

Built upon from Chaffey and Smith (2009).

Assessment advice

It would be a good starting point to consider how your might use the SOSTAC® model early in your assignment planning for this module. You could use the table above as a checklist to ensure you have added sufficient detail to your own plan.

2 The e-marketplace

The e-marketplace is continuing to grow rapidly and seamlessly across borders. For some people the *electronic marketspace,* already inhabited by over a million customers, (Chaffey and Smith, 2008) in its extreme, even goes as far as becoming a virtual life (hence the development of sites such as Second Life and other highly immerse-based virtual communities). For the majority of consumers, the e-marketplace is likely to remain less intense than the 'surreal' world of virtual lives, robots, info-bots, shopping-bots (such as fridges which can automatically re-order products), portals and infomediaries but nevertheless it is a big part of the future of all businesses and a large majority of consumers.

Already highly complex technologies exist in our everyday world often without our knowledge. Through a powerful combination of computer chips, cordless or wireless technology, high speed data transfer protocols, high speed data access, WAP (wireless application protocol) and Bluetooth, technology more advanced in computing power than that used for the first moon landing are now highly integrated into modern life (Chaffey and Smith, 2008).

Global case study

A consortium including Rolls Royce are currently working on a project titled WiTNESS project which is partly government funded and designed to improve air safety.

A wireless sensor system will enable the real-time monitoring of critical components during flight so that engineers can be on standby when the flight lands or rectify where possible remotely. The system will be used to help identify technical faults, optimise performance and monitor the overall health of the aircraft. Wireless is said to be a crucial capability to reduce the costs associated with wired sensor cables.

The system will give aircraft operators the ability to detect and rectify problems before they lead to serious consequences so ultimately, these systems could make a significant impact on aircraft safety.

Work on the first prototype wireless sensing system began in September 2009, and the consortium plans to deliver a range of commercial application demonstrators by the end of 2011.

Adapted from Walko (2009).

The range of product and service opportunities as a result of these advanced technologies is therefore highly exciting and not likely to diminish soon.

2.1 The markets

Traditionally the bulk of Internet transactions are between the following parties:

- **Business to business** (B2B) – which is the larger category of Internet usage with ten times more revenue (Chaffey and Smith, 2007)

- **Business to consumer** (B2C) – this is the area you will be familiar with as an Internet consumer!

Vertical and horizontal e-marketplaces provide online access to businesses within the supply chain and even between sectors.

Horizontal e-marketplaces connect buyers and sellers across industries or regions. VertMarkets (www.vertmarkets.com) provides industry specific online sales and marketing products to small and medium sized organisations. The trading platform brings together buyers and sellers in over 60 industries to create markt opportunities for businesses in in IT, public sector, life sciences and food industries.

GlobalNetXchange is another trading hub which acts as a global e-business consortium for large retailers including sears, Carrefour, Coles, Myer Ltd and J Sainsbury. The hub connects retailers, manufacturers and their trading partners to reduce costs and improve efficiency by streamlining and automating sourcing and supply chain processes.

2.2 Online behaviour

According to e-consultancy (2009) in their annual Internet statistic compendium, the following facts outline the extent of the e-marketplace and issues that require careful consideration:

- More than 85% of the world's Internet users surveyed have purchased something online.

- 80% of consumers say they would be less likely to return to a retailer's website after a negative online shopping experience.

- Nearly 40% said a frustrating online experience would make them less likely to shop at that retailer's high street store.

- 91% of B2B Technology decision makers surveyed by Forrester said they were spectators in some way of social media (blogs, user-generated video, etc). Only 5% said they were non-participants.

When assessing how consumers behave online segmentation is equally important. Particularly in relation to their propensity to initially go online, interact and then make online purchases. Issues which may determine how consumers behave will be dependent on:

- **Type of access to the Internet** – dial up, broadband, mobile access, personal or work related

- Extent consumers are **influenced** by online channels – as part of the research process at the decision making stage, in terms of their online activities

- **Propensity to make transactions online** – have no need, prefer to shop in person or have security concerns

- **Demographic characteristics** – rapidly changing compared to early Internet usage, differences are more apparent in the purpose of Internet usage

- **Usage level** in terms of frequency of use and purpose of use can help profile customer segments psychographically

Think about your own online behaviour – how would you categorise yourself?

Organisations vary in their demand for online services according to:

- size of the company
- industry sector and products
- organisation type
- application of support services
- country and region

Guerrilla marketing expert Jay Conrad Levinson has a number of articles on his website. One of which refers to research conducted by the Pew Research Centre into online habits and behaviours. From this Levinson states that people have a goal orientated approach to using the Internet and as such concludes that:

"What does this (the research findings) mean to e-marketers? It means that if you're constructing a site for goal-oriented consumers, you'd better make sure you can help facilitate their seeking. Rather than focus on entertainment, flash, and useless splash screens, the most effective sites are those that help people get the information they want when they need it. Straightforward data, information that invites comparison, and straight talk are going to win the day."

To read the full article and access others discussing technology and guerrilla marketing generally visit:
http://www.gmarketing.com/articles/read/15/What_Do_People_Want_Online?.html

There are many articles on the site worth reading the following are particularly valuable for this chapter:

http://www.gmarketing.com/articles/read/116/Guerrilla_Marketing_Online.html

http://www.gmarketing.com/articles/read/101/Online_Conference_Credibility.html

http://www.gmarketing.com/articles/read/98/Internet_Marketers_Missing_the_Boat.html

3 e-marketing mix

The marketing mix is a well established concept, we briefly looked at the beginning of the chapter at how the mix would be influenced by digital marketing. It is now time to look at each of the 7Ps in turn in terms of how they need to be changed in order to accommodate the needs of online consumers.

3.1 Product

According to Chaffey et al (2009) there are a number of aspects within the product element of the mix which requires alteration.

3.1.1 Options for varying the core product

Adding digital value according to Ghosh (1998) is essential for digital products, key questions to ask include:

- Can I offer **additional information or transaction services** to the existing customer base? eg Amazon offering 'look inside the cover'?

- Can I address the **needs of new customer segments by repackaging current information** assets? eg Amazon offering DVD rental

- Can I use my ability to **attract customers to generate new sources of revenue** eg Amazon up-selling consumers to Prime accounts with *free* next day delivery?

- **Will my current business be significantly harmed by other companies** providing some of the value I currently offer? – the strength of the Amazon brand built through high levels of consumer experience with exceptional service levels help to protect from this threat.

3.1.2 Options for offering digital products

Publishing, TV and other media related companies can offer digital products such as online newspapers, videos, additional digital content in a flexible manner and often at differentiated prices. Marketing Week for example traditionally had its paper version of the industry journal, then moved online with free limited content and full archive access for subscribers.

3.1.3 Options for changing the extended product

The after-sales and customer service elements of the extended products can be maintained online, Chaffey and Smith (2008) include the following possible tools:

- endorsements

- awards

- testimonies

- customer lists

- customer comments

- warranties

- guarantees

- money back offers

- customer service elements – some sites offer the option for an online chat with a service representative or a request for a telephone call-back

- assistance with selecting the appropriate product eg gift buying guides, skin type analysis for cosmetics.

3.1.4 Velocity of new product development

NPD can be sped up as a result of access to online research tools. Partnerships are also enhanced by being able to work online with collaborative partners and their network of agencies looking to jointly launch products.

When Colgate were developing a new child's toothbrush, they were able to complete all the necessary prototype testing in less than a day with children online selecting their preferred colours and overall design. This was more efficient due to the speed but also had fewer resource implications than traditional testing.

3.1.5 Velocity of product diffusion

Word of mouth communication is highly influential on the rate of diffusion of new products. With the speed and culture of sharing opinions online, the Internet has intensified this powerful medium.

The following play a significant role:

- viral marketing

- blogs

- product reviews either on standalone sites such as Revoo or increasingly on retailer sites eg Amazon

- social networks

3.2 Price

The Internet has turned the concept of pricing upside down. Chaffey and Smith (2008) report that increasingly situations occur where those who are paying for online services one day are actually paid to use the service the next. The authors cite AOL as an example. Initially AOL paid ABC News for content but now ABC News pay AOL to place content on its pages. In addition, consumers are sometimes paid to watch ads and participate in online research.

Options for differentiated pricing include:

- **Subscriptions** – eg 3 months, 6 months etc

- **Pay-per view** – fee to download a single file, e-book, podcast, film

- **Bundling** – grouped content at reduced prices. Econsultancy for example sells individual reports on a pay per view basis or a bundled subscription for access to all reports.

- **Ad supported content** – there is no charge for content as sites are funded by revenue from advertisers – the free newspapers of the online world!

Prices are also under downward pressure for a number of reasons:

- **Increased competition** with an increased availability of global suppliers (when consumers search more widely online)

- **Reduced margins** for online companies as a result of more efficient web-enabling databases and processes

- **Commodotisation** as a result of reverse auctions for both consumer and business products

- **Price transparency** due to publishing price breakdowns on the web and price comparison websites

- **Disintermediation** where consumers are able to cut out brokers and deal direct with manufacturers and service providers

- **Leasing is increasingly popular** moving consumers from fixed prices to rental options even in products such as designer handbags

3.3 Place

Increasing the opportunities for consumers to purchase goods and services online by maximising presence where the target audience are likely to visit is a key strategy.

Clothing manufacturer and online and catalogue retailer Boden not only promotes its clothing range in traditional media but runs many banner ads on sites with direct click through links to relevant pages on its site. The sites are selected as affiliate partners because they are frequently used by Boden's typical profile customers.

Chaffey and Smith (2008) identify a number of key changes for distribution channels. These include:

- **Disintermediation** – as we have already mentioned with respect to pricing

- **Reintermediation** – an emergence of new types of online broker middlemen uniting buyers and sellers eg Bizrate, Comparethemarket etc

- **Infomediation** – middlemen hold information to benefit customers eg Comparethemarket etc

- **Channel confluence** – distribution channels start to offer the same deal to the end customer

- **Peer to peer services** – music and video swapping services provide an entirely new distribution Model

- **Affiliation** – affiliate programmes turn customers into sales people. Amazon uses this to great advantage where they establish affiliate partners who are able to sell products via their own website through Amazon.

3.4 Promotion

We have already covered promotional tools in earlier sections of the chapter and it is a topic which is covered extensively in Chapter 6 (plus later chapters when we move to look at integrated communication strategies). It is worth us noting here however that e-marketers need to mix the optimal promotional mix using a range of online promotional tools.

Online promotional challenges can be summarised in six key issues:

- **Mix** – using the most cost effective online tools to acquire target customers

- **Integration** – Online and offline communication must be integrated in order to maintain a consistent brand message

- **Creativity** – many creative opportunities are presented by the Internet. Virtual immersion enables consumers to become avatars in another 'world'

- **Interaction** – enables consumers to interact with one another and with brands through games etc. We will return to the related concept of 'sizzle' in Chapter 7.

- **Globalisation** – websites are open to most of the world, localised branding and strategy is therefore not valid when it comes to digital strategies

- **Resourcing** – resources to design and maintain content along with the related customer service and fulfilment requires significant resource.

3.5 People

As with services marketing, for online marketing success staff are critical. The following uses of people (staff) within the online service delivery or act in the capacity of a person (Chaffey et al, 2009):

- **Autoresponders** – an automatically generated response
- **Email notification** – order status and delivery details are provided to reassure and inform customers
- **Call-back** – by submitting a phone number a call centre rep calls customers to speak in person
- **Live chat** – instant messaging enables live chat between sales or customer support agents and
- **Frequently asked questions** – can pre-empt common questions and negate the need for staff contact
- **Virtual assistants** – guide through purchase choices
- **Co-browsing** – call centre operators are able to view the customer's screen in order to improve the quality of advice they provide
- **Customer reviewers** – customers recruited to provide comments for other consumers. Advocates of the brand are the most positive and sought after consumers to take on this role due to the positive benefits of word of mouth

3.6 Process

The actual process of transactions and how smoothly customers can navigate through websites along with the process of any after-sales service are critical to online marketers. It is through these processes that customers will judge the quality of the brand.

Customers who experience problems with order processing are highly unlikely to return to a site as this will form a significant 'moment of truth'. Likewise, if online transactions are effortless but then delivery is delayed, loyalty is likely to be diminished.

3.7 Physical evidence

The qualities associated with the website itself will help to build the physical evidence aspects of the brand. The higher the perceived quality of the website, the higher the levels of trust from consumers.

Global case study

Basically Black is an online retailer of women's clothing. Their USP is that they only sell black clothing because they have found that 40% of a typical UK woman's wardrobe is made up of black items.

The company epitomises personal levels of service and attention to detail. It is also an exemplary example of how many small organisations have begun to trade as a result of the increase in online retailing.

Read the 'About us' section of the website at:

http://www.basicallyblack.com/pages/about.aspx

In a moment of author indulgence, I will share with you a personal anecdote of my own experience of using the company. I will leave it to you to consider how likely I am to use this retailer again!

In December 2009 we launched the Global Marketer Programme in Kuala Lumpur, Malaysia. Knowing that the temperature in KL would be significantly higher than in London I decided that I needed a new business suit to wear at the launch event. The problem was that clothes in the shops in the UK were geared for the cold winter season. As part of my information search, I used Google to search for linen suits and came across the Basically Black website.

I ordered a suit online and opted for next day delivery (paying a supplement) because I had left my purchase a little late and was due to fly in a few days' time. Twenty minutes after I submitted my order, Janey, the owner/founder of the company called just to check that I had meant to select the next day option because as it was a Friday, it meant that my delivery wouldn't arrive until the Monday. In a brief conversation I confirmed that yes, I had intended to opt for a faster delivery and explained why I needed the items so quickly. Janey mentioned that she had a number of 'summer sale' items in her office and if I needed anything else then to call her.

When my order arrived on the Monday, it was beautifully wrapped, with a hand tied ribbon and tissue paper (hence the physical evidence was highly reassuring) with a note wishing us luck with the programme launch.

This example demonstrates the following:

- Personal service from helpful staff is highly valued by customers

- Attention to customer's needs can really help future sales

- Seamless transactions with all elements of the process working effectively generate customer satisfaction

- Segmentation and well considered USPs are equally valid to online retailers

- Exceptional levels of customer service are key to loyalty

- The 'About You' section should demonstrate how building a good brand/corporate story will help establish trust and empathy with the brand

You have probably already guessed that yes, I am likely to order again.

Activity 4

Complete an audit of your organisation's online marketing mix.

1 Discuss how digital marketing integrates with the wider marketing plan

- Technological developments associated with digital marketing
- SOSTAC® and planning for digital marketing strategies.

2 Describe the e-marketplace and discuss things to consider with regard to online consumers

- B2C and B2B
- When assessing how consumers behave online segmentation is equally important
- Business usage depends on a number of diverse factors
- Significant shifts in the marketing mix are occurring as a result of the Internet
- Online and offline mixes need integrating

3 Evaluate the benefits of digital marketing

- A valuable route to market
- Opening of competitive markets to smaller operations
- Enhanced communication opportunities
- New services at reduced cost
- Valuable promotional and research opportunities

1   This will depend on your own research within your own organisation.

2   This will depend on your organisation.

3   This depends on your own online behaviour.

4   This activity will take you some time to complete. It is worth doing as it will directly help with your assignment prep work.

Brand Week (2009). Survey: CMOs Not Happy with Digital, Brand Week, 18th April 2009. Available online at: http://www.brandweek.com/bw/content_display/news-and-features/digital/e3if46ca983d59bcb8f9823719f0e98475c?pn=2 [Accessed 27.1.09].

Brassington F. & Pettitt, S., (2003). *Principles of Marketing*, 3rd edition. London: FT Press.

Chaffey, D., Smith, P.R., (2008). *Emarketing Excellence.* 3rd edition. Oxford: Butterworth-Heinemann

Chaffey, D. et al, (2009). *Internet Marketing: Strategy, Implementation and Practice.* 4th edition. Harlow: FT Prentice Hall.

Econsultancy (2009) Internet statistics compendium, December 2009. Available online at: http://econsultancy.com/reports/internet-statistics-compendium [Accessed 3.12.09].

Ghosh, S., (1998). Making Business Sense of the Internet. *Harvard Business Review*, March – April p. 127.

Walko, J., (2009). *Consortium to Develop Wireless Sensor Based Aircraft Systems*. EE Times. Available online at: http://eetimes.eu/uk/219400385 [Accessed 26.11.09].

Jobber (2007) Jobber, D., (2007). *Principles and Practice of Marketing*. 5th edition. Maidenhead, Berks: McGraw Hill Education.

Levinson, J., (2009). Guerrilla Marketing Association [Online blog]. Available at: http://www.gmarketing.com [Accessed 26.11.09].

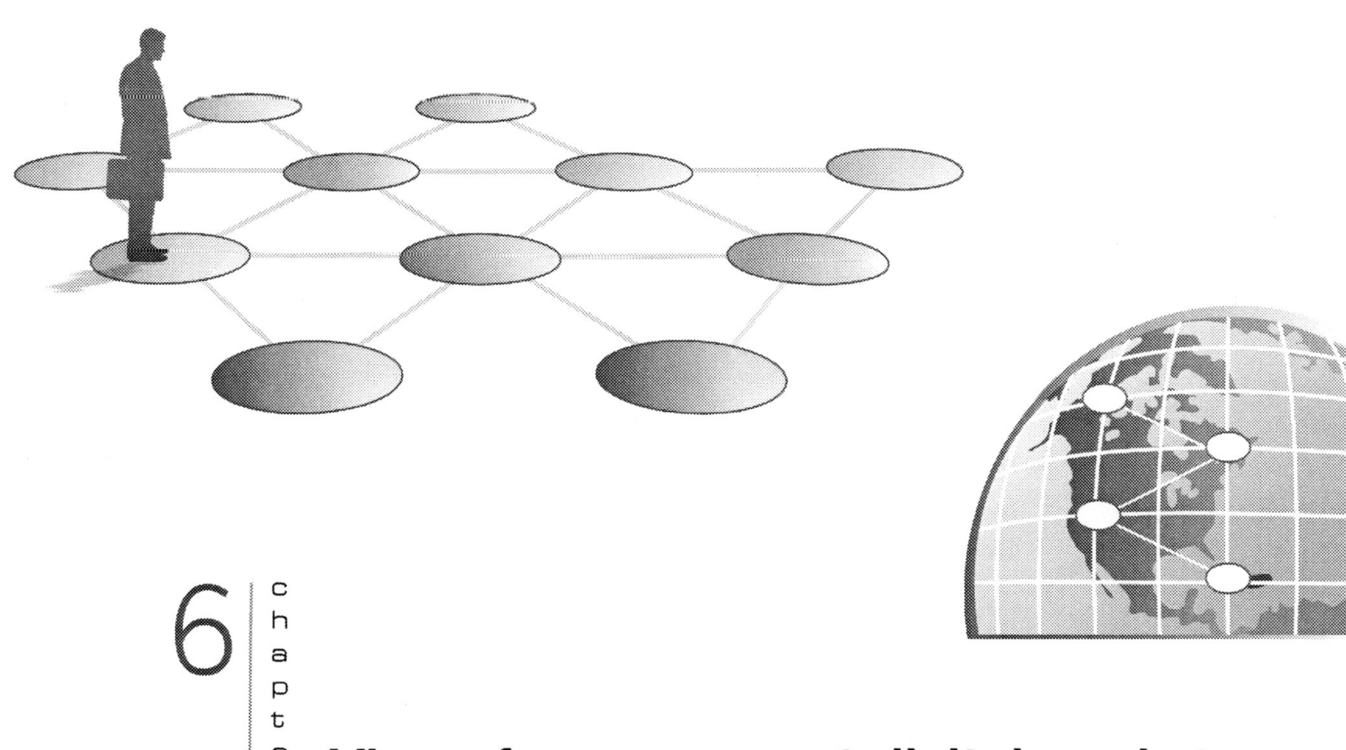

6 | chapter

Views from an expert digital marketer

This chapter is very different to any others within this Study Text because it focuses exclusively on the practitioner view. Digital marketing is changing so rapidly, that it is one area where theory can often lag behind practice. We have taken the opportunity to learn from one of the leading experts in this area and bring you our practitioner expert chapter.

Damon Segal is a GMN Advisory Council member and has contributed to the development of the overall Global Marketer Programme. Damon is a digital marketer, professional speaker & CEO of Action Graphics International. He has been a major force in the design and marketing arena for 20 years.

Having worked for businesses ranging from luxury retail and toys to celebrity and financial sectors, Damon has a variety of approaches to practically solve all marketing hurdles. Damon regularly speaks at international conferences on the subject of digital marketing.

Given the alternative nature of this chapter, we provide assessment tips throughout and reflections from Damon as shown in the GMN viewpoint feature. There are no formal activities and debriefs however.

The chapter provides a balance between the theoretical underpinnings of the previous and subsequent chapter.

Contents

Chapter learning outcomes

By the end of this chapter you will be able to:

* Discuss in detail a range of practitioner based digital marketing issues.

1 Introduction

It was 1957 when the United States formed the Advanced Research Projects Agency (ARPA) within the Department of Defense. It was to establish the US as the leader in military science and technology. It was this project that lead to Tim Berners-Lee publishing the first web page on the WorldWideWeb on Christmas 1990, and on October 24, 1995, the Federal Networking Council (FNC) unanimously passed a resolution defining the term Internet.

The Internet Society (ISOC), *Histories of the Internet*
http://www.isoc.org/internet/history/ [Accessed 1 November 2008]

This has developed into the Internet we know today where billions of web pages on every subject are available, we have Internet telephones, Internet music, Internet television, auctions, shops and, of course, email. We started with crude telephone-connected modems at connection speeds of 9,600 kbps and now we have fast broadband. Once you could only access the web from your desktop PC with a wire connected to the wall, now you can access it from the top of a mountain via a hand-held mobile device. This means that from a marketing point of view we can reach people wherever and whenever we want. Not only this but it means we can stay in touch with individuals, targeting content specifically for them whether they are at work, at home or in transit using a mobile device.

World Internet usage and population statistics

World Regions	Population (2008 Est)	Internet Users Dec/31, 2000	Internet Usage, Latest Data	% Population (Penetration)	Usage % of World	Usage Growth 2000-2008
Africa	955,206,348	4,514,400	51,065,630	5.3%	3.5%	1,031.2%
Asia	3,776,181,949	114,304,000	578,538,257	15.3%	37.5%	406.1%
Europe	800,401,065	105,096,093	384,633,765	48.1%	26.3%	266.0%
Middle East	197,090,443	3,284,800	41,939,200	21.3%	2.9%	1,176.8%
North America	337,167,248	108,096,800	248,241,969	73.6%	17.0%	129.6%
Latin America/Caribbean	576,091,673	18,068,919	139,009,209	24.1%	9.5%	669.3%
Oceania/Australia	33,981,562	7,620,480	20,204,331	59.5%	1.4%	165.1%
World Total	**6,676,120,288**	**360,985,492**	**1,463,632,361**	**21.9%**	**100.0%**	**305.0%**

As you can see from the information above, Europe, North America and Oceania/Australia have a significant reach based on usage figures. In the United Kingdom over 68% of the population are Internet users (41,817,847 Internet users in June/08, 68.6 % penetration, per Nielsen Net//, 14,361,816 broadband subscribers as of June/07, per OECD, 23.6% p.r.) (Miniwatts, 2008)

With this kind of penetration marketing on the Internet has become one of the most measurable and cost effective methods of creating brand awareness and driving sales. In the UK £2.8bn was spent on Internet advertising in 2007, an increase of 38% on the previous year. This spend represents 15.3% of the total money spent on advertising in the UK. For the first half of 2008 spend was up 21% at £1.7bn from Jan-June. As the Internet embeds itself more into our lives it is expected that Internet advertising will outgrow television revenue by 2010.

Internet Advertising Bureau UK and PwC. http://www.onlinecontentguide.co.uk/news-2009/2.8-billion-spent-on-internet-advertising-in-2007.html (cited Written by Jack Wallington, Monday, 07 April 2008, *£2.8 billion spent on internet advertising in 2007*) [Accessed 12 November 2008]

IABUK. Available at: http://www.iabuk.net/en/1/iabknowledgebankadspend.html [Accessed 2 January 2009]

2 Key concepts for using the Internet as a marketing tool

(a) **Defining requirements for a website**

Sometimes the answer is as clear as day, sometimes it is lost in the depths of peer pressure and mysteries. Many businesses will already have a website, but many businesses will be unaware of how that website affects their business.

(b) **Different types of websites**

There are many different types ranging from content websites to e-commerce. Understanding which is right for a business is a critical step to achieving success on the Internet.

(c) **Importance of usability**

Usability is the ease with which users can operate your website. There is also an accessibility issue regarding usability for people with disabilities.

(d) **How to get a website built**

Choosing an agency is no simple task. There are over 11 million results on Google.co.uk when searching for web design agency, even on a phrase match search "web design agency" there are over 300,000 results. Narrow this down with "web design agency london" and you end up with over 4,000 results. Choosing the right team is critical to the site's success.

(e) **Hosting**

Web hosting is where your website is kept. A website is held on a computer (usually a server) that is connected to the Internet. Your website is a collection of computer files, the server is set up in a way that allows people entering your web address (URL, or Uniform Resource Locator) to view your site.

3 Defining requirements for a website

To analyse this you must first understand what purpose each type of website can fulfill. There are a number of different types of websites which we shall refer to in section 4. While reading this bear in mind Chaffey and Smith's (2008) notion that you MUST at all times consider the purpose or objective of the website. The authors provide possible answers to the question *"How can my website help my customers?"*

- Help them buy something they need
- Help them find information
- Help them save money and time
- Help them talk to the organisation
- Help them enjoy a better web experience

Chaffey & Smith, (2008 p.223).

4 About different types of websites

4.1 Content websites

These sites are primarily online brochures or information services. They provide content on a business (we should not ignore the fact that sites also exist for organisations, charities and many other categories). Bearing in mind the Internet is often the first port of call for people researching a business or looking up a subject, these kinds of sites are the most popular on the web. A site like this is there to create a window on what a business does; if it does not represent that business correctly then it is probably doing more damage than it is good. Websites give credibility, confidence and substance to a business, but equally they can destroy it.

Having a content site is a minimum requirement nowadays. There are some businesses that are so large and exclusive they do not need an Internet presence but for the rest sites provide great leverage in obtaining new customers and impressing suppliers. You probably sell something in your business and what better way can you get your brand in front of your customer? Any time of day, 365 days a year, your message is accessible from pretty much anywhere on the globe.

Research published on BBC News said the brain makes decisions within just one-twentieth of a second of opening a website. If the user decides a site is good, they will try to justify that feeling. This makes them quite forgiving if other areas of the site do not prove quite as good as the first impression. If the first impression is less favourable, then users will quickly be off your site and on to your competitors' sites. It is important that your site is visually right for your industry and that the visual impact is a good one (BBC, 2006).

Content sites used to be built in HTML, the basic language used to create web pages. These pages are static and cannot easily be changed except by someone with web building experience. The other style is content-managed websites, which usually give the owner access via an admin/management control panel. They feature an editing facility that lets you manage both text and images as they appear on the site. Content management systems (CMS) come in many forms. Some are very complex to use and can cause the user to require substantial amounts of IT knowledge. Others are simpler and often more effective. CMS systems are readily available but my advice is take a good look at your options before you buy. It is usually preferential to have a CMS controlled site as it puts you in the driving seat. You can modify and control your site instantly without having to rely on an agency or one man band to get your change done immediately. There are also the financial savings to consider.

> ### GMN viewpoint
>
> If a business is not on the web then the competition will be, if a site visually lets a business down the competition will be there when they move to the next site and if the content is weak the competition will...well, enough said. The most basic start to getting a business on the web is the creation of a good content website. Think carefully about what you need to say and why you want to say it. Remember the basic marketing principle of what benefit can a business offer the user. Don't go into lengthy histories of a business, as people don't often spend that long on a website – so get to the point fast and keep it clear.

4.2 Web based application web sites (Webapps)

This is commonly referred to as Web 2.0, a website that runs as an application with useful functionality. Applications can be anything from an online polling system to a site that facilitates the gathering, matching or distribution of information. Sites like linkedin.com or monster.com are great examples of this. Electronic and software companies often use webapps to provide customers with support and update downloads. Webapps are pieces of software that can be accessed via the Internet or an intranet. It means a business can have a centrally modifiable application that can be accessed via a browser interface by multiple users in multiple locations.

If you are considering updating or creating a new site, webapps can streamline simple or complex administrative tasks, allowing staff resources to be saved. For example, imagine you have a business that manufactures coffee machines. Nearly every day you receive calls asking for copies of instruction books, people looking for spare parts and those all too common requests like "Where can I buy coffee pods to fit my machine?" Let's say you get 12 of these calls a day; at five minutes a call that is one hour of staff time per day, 20 hours a month and 240 hours a year. How much does an hour of that staff time cost the company? Let's say a conservative £30 with overheads, so those calls actually cost your bottom line over £7,200!

Imagine if that business's telephone hold system told people to look at the website for this information, so a user could simply select their coffee machine from a list then download a PDF (Portable Document Format) of the instruction book. Imagine if that business had an e-commerce system selling those all-important spare parts. Users could simply select a list of accessories or parts for a specific machine, pay online for them, then the orders could go straight to a distribution centre. The business could even sell the coffee pods online itself or direct the customer to a local retailer. If the webapp is written well, you can manage changes to this information in-house with ease.

Webapps can also facilitate collaboration between communities and groups with websites allowing users to add and edit content. These can take the shape of web forums or wikis, the best known being Wikipedia.

There are many programming languages and many platforms and browsers. Developing web applications that are compatible across all these platforms is no easy task, so you should never under-estimate the amount of time it will take to create these systems. Some applications can be built using existing technologies like Google Maps or YouTube. You can use APIs (Application Programming Interfaces) to work with these technologies from your own site. SDKs (Software Development Kits) are also available for many technologies such as Sage or Adobe's Flex. SDKs provide a way for software engineers to develop applications that work with these existing softwares.

The concept of webapps is endless and limited only by your imagination and your developer's technical skills.

GMN viewpoint

Consider if there are tasks or processes in a business that could be streamlined or automated via a website. Consider how a website can add value to a customer or community. Webapps, although more costly to develop, can save huge amounts of time in your business and both save and generate money if carefully planned. One very important factor in the creation of a web application is preparation. All information and specification should be clearly defined before any coding is begun. Webapps typically take from 3 months to 12 months depending on complexity and the size of the development team.

4.3 E-commerce web sites

A company doesn't have to have a shop selling widgets in order to do business online.

E-commerce sites can sell tangible products or intangible services. You could be selling toys or ebooks, insurance or tickets. Credit card transactions are taken over a secure connection known as HTTPS. A website can be made secure by applying a special registered certificate called an SSL (Secure Sockets Layer) capable of high levels of encryption, providing secure communications on the Internet. A certificate can be bought from companies like Thawte.com, Instantssl.com or VeriSign.com. Cost varies from £30 up to low thousands, depending on what level of security a business requires.

There is a way around this expense, which is to use a payment gateway. These are facilitators for e-commerce payments where payments take place on the provider's own secure servers. This means your site does not process the actual payment process. Transactions are done in real time although different gateways have different terms and charges. Paypal is really easy to set up and other common gateways are Securetrading.net, Worldpay.com, Protx.com, Authorize.net and Secpay.com.

Think how an automated payment system would cut down on the administrative time taking calls and orders and increase the ability to sell to more people.

E-commerce facilities are there selling every hour of every day, they can take multiple sales at the same time and aid in creating impulse sales by presenting associated products or services as part of the sales process. If a business is heavily orientated to up-sell it is important to ensure it has an e-commerce solution able to facilitate this.

Many SME businesses commission an e-commerce site and expect to get rich overnight! This is often not the case, and below is a real world analogy.

You decide to open a shop selling clothes, you take a lease on a lovely retail space just off a high street. You spend £20,000 making it look pretty and inviting, and you stock it full of wonderful clothes. In the end you sit on your stool and admire your handy work... What now... You suddenly realise you forgot to budget for marketing and advertising, your budget is blown on stock and fixtures! Without telling people where your shop is, how will they know you are there?

This is the most common error made in setting up an e-commerce shop. Search engines are great but a new domain name may fall foul of the 'sandbox theory', a holding area for new sites before they are released into the Google index. Google has publicly denied that this exists and my experience has shown some sites get indexed on release; but primarily they have been sites that have a great deal of content and have been built with SEO (Search Engine Optimisation) at the core of their development. Even then it takes time for the site to move up the ranking. If your business is opting for an e-commerce website make sure you have a strategy and a budget for when the site is complete. The more money there is for marketing, the more traffic that can be bought for the site. The challenge is making sure the business has a product to sell that will fetch margins and profits high enough to cover the cost of customer acquisitions.

Another common error is to forget that customers often buy more than once on a site so if an average profit per sale is £25 you have to bear in mind that customer may buy ten times a year, resulting in £250 from that customer. This makes acquiring a customer a bit more affordable.

 Developing Integrated Communications Strategy

4.4 Social Network sites

Facebook, MySpace, Friends Reunited, Yahoo! 360°, Ecademy, Linkedin and Xing are social networks that connect people at low or zero cost. They facilitate online networking, allowing people to expand their personal and business contact base. Creating a social network site can be great if you have a specific area to focus on. Linkedin.com is a powerful business tool, allowing professionals to share their business network with colleagues and friends. It allows questions to be posed to complete networks and answers to be given quickly via its inbuilt systems. As business tools they can be great, but as social tools they are incredible – just ask anyone under thirty who works in an office that does not restrict access to Facebook! In an article by the BBC – Facebook 'costs businesses dear' – it reports a study has been made showing Facebook could be costing businesses £130m a day! But think how much money it is generating for the private owners of Facebook. How would you like to own a Social Network site now?

If a business is thinking of creating a Social Network site consider what is out there already, research who is doing it well and how they did it. *Inside Facebook: Life, Work and Visions of Greatness* is available on Amazon and might be a good starting point. You will need a great value proposition and a clear vision of where you want the site to be positioned. It is the classic "start with the end picture and work backwards" scenario. A social network site needs to connect people so consider who you want to connect and why. Members are the key to a social network's success so make sure your plan is oriented to obtaining them. Making membership free to start with and focussing on a strategy that builds page views for advertising is a good route, the only issue with this is it takes a while before a business can see a return. Many social network sites introduce membership fees after they are established before users can access more premium content on the site.

4.5 Blog sites

A Blog site is a collection of thoughts, articles or notes made by individuals or organisations to keep users informed. A Blog site can be as simple as a daily diary of a traveller to the official blog for Google (www.googleblog.blogspot.com) where you can get a look at life inside Google as well as discover news, new products and events of interest. Google has over 100 official blogs.

Blogs are an excellent way to keep in contact with large groups of people. There are several Blog systems on the web that are free, with blogger.com and wordpress.com the most notable.

A Blog often forms part of a Social Media Optimisation Campaign.

4.6 Directory web sites

If you talk to someone who wants a directory site you can see the £ signs flashing in their eyes. It usually goes along the lines of... 'if I charge £199 per year to be listed on my directory I only need a few hundred people to be making over £60,000 a year without having to do any work!' It is the words 'only need' and 'without having to do any work' that you need to be careful of.

Directories are great and they do provide an excellent service and if you get it right, people will pay good money to be on a directory. First, let's look at why someone might advertise in a directory. A specialist directory of businesses which provide services to a specific industry would attract advertisers. For example, where would you look for suppliers to the aerospace industry? If you check in Google you will find Business.com with a dedicated directory listing for this industry. With directories you have a bit of a horse before the cart problem. Some of the keys to a successful directory are how people will find it on the Internet, how many people are listed on the site already and how easy is it to find what you are looking for once you are on the site.

Let's look at the example of the aerospace suppliers directory... if you were asking someone in this industry to take a listing on your new specialist directory, the obvious question they will ask is how many people look at this directory each month? It's a new directory so this will be a tough one to answer as you probably have no proper statistics. The next question is where is the site advertised? You had better have a good answer for this one. Search engines are unlikely to have you listed on top pages that quickly so don't rely on this to save you. Paying to advertise your directory in trade publications, whether online or in print, is a good starting point. Finally, when looking at the site and seeing you have only 15 companies listed does not inspire the person on the end of the line to buy a listing on your site. So what can you do about this? The trick is getting people on to the directory by making it valuable for them to be there.

Example of small directory startup

Start with an offer, say, the first 100 people get to sign up for £49 for 3 months and they get 3 months added on free! Also tell them that their time won't start until you hit the 100 people mark and that you will have an official launch which will be covered in a great press release. This makes the proposition a fairly easy sell as it is not too expensive and they get something for nothing. The next question is where are you going to advertise? Well, you know you are not going to launch till you have 100 business listings so you know you will have £4,900 to play with over a six month period (I say six months because this is how long you have until your 100 businesses decide if they want to renew their listing). Don't expect to make any profit at this point. The money can be committed to a small advert in a trade magazine each month and/or Google pay per click. So when you get asked that question you can say you will be advertising on launch in XYZ Specialist magazine and on Google. Now £49 for six months sounds even better. The "how many people?" question is now answered with "We are launching once we reach 100 businesses (could be 500) and you get listed free until we hit that number, all you have to do is commit to £49 for a three month listing with three months extra thrown in free!" Once you launch you have great answers to all the main questions. "How many visits?" still might be a challenge, but you have a great answer for where you are advertising, and when people look at the site you have 100 listings so it looks good. Finally, the price: keep it keen and the decision is a simple one. When you have thousands of listings, start adding value to the site and charging more.

An established company launching a directory would be slightly different as budgets are usually higher, but the principle is the same: make the offer to be listed irresistible and remember that people want to know the directory is being found, so have a great marketing plan.

There are many other models, including making the directory free and generating profit via advertising.

GMN viewpoint

Directories can be great revenue generators. But getting them started is HARD. Remember to check out the competition, after all I found a directory of aerospace suppliers on page one of Google. Find out what the competition is doing and do it better. Once you're up and running, make sure you keep up the momentum.

4.7 Intranet sites

Intranets are websites that are only accessible to private users across a private network. They are excellent tools for organising and storing information that is relevant to the operation of a business. Intranets provide an excellent way to communicate with staff, helping to increase productivity and spread a corporate message to all levels. A few good uses might be giving staff access to a company handbook, providing processes, quality control forms, human resource information, applications for holiday time, contact databases, downloadable files, brand guidelines etc.

If you are going to create an intranet, first think what purpose the system has, what the business wants to achieve from having an intranet. Security is a big issue, so consider how the site will be protected and what protocols would be in place if someone left an organisation, often an area that is forgotten. Will the information be managed from a central place or will multiple users have the ability to control content on the site, even the ability to collaborate with information?

GMN viewpoint

If you are looking for a way to communicate or collect information from across an organisation, an intranet is an excellent tool. Keep in mind the security aspect and consider which areas of your business can be streamlined by introducing a system like this.

4.8 Extranet sites

An extranet is the same as an intranet but is also available over the Internet. It is still secure for private users only but it can be accessed from outside of a local network. This means that you can share and collect information from multiple locations.

GMN viewpoint

Security is an even bigger issue, so make sure a good plan for this is in place.

There are many others types of sites such as web portals, gaming sites, warez sites and more.

5 Importance of usability

Often when your website is being built the people closest to it don't see how difficult the site might be to use.

Another real world analogy

It's like having an old car. If you are the owner of that car you might know to lift the door a little when opening it, then to pump the accelerator twice before turning the ignition. Once the car starts you rev to 5000 rpm for three seconds and let it come down to a nice idle speed or it stalls. Now think what it would be like to give someone the keys to your car and tell them to go for a drive. They might not even get the door open! Some websites are like this and usability is often a key in getting conversions (sales, enquiries, downloads, etc) from a site.

Usability is the ease with which users can operate your website. There is also an accessibility issue regarding usability for people with disabilities (see http://w3.org/WAI/ for details). An important factor is make your content to the point and don't try to be too clever. If your shop has a basket, call it a basket not a trolley or some other obscure title. Make sure site areas or groups are clearly defined and recognisable. Be consistent in layout, navigation and processes. Give people access to all parts of the site easily and guide them to achieve your site's goal by having defined paths through your website. Consider the position of elements on a page and remember to put important things in the top left. Most of us read from left to right so these will be seen quickly.

Three-Click Principle: It is suggested that nothing useful on your site should take more than three clicks to achieve or reach. If you have articles that you want people to read, or products you want people to buy, you should be able to do this in three clicks. A shopping site should allow you one click to find a product, one click to add to basket and one click to checkout, for instance.

Twenty Dos and Don'ts of web design

1   If your organisation's branding is in the top left corner of a page, users will expect this to link back to the site's home page.

2   About Us commonly refers to the organisation's information page.

3   Navigation must remain constant on every page and is either across the top or on the left side, sometimes both.

4   Contact Us is found as the last item on a menu.

5   Anything you display that is animated and appears above a logo will be considered an advert by users.

6   Don't make the fonts too small and allow users to increase font sizes.

7   Make links obvious and differentiate between visited and non-visited links.

8   Don't open new windows on a click.

9   Use minimal flash or animation, as users equate animated content with useless content.

10   Write for the web, make it concise and to the point as it's better for search engines and it's better for users.

11   If you have a search facility on your site ensure it gives good results, not just huge lists of irrelevant content.

12   Make your site compatible with multiple browsers used on the web. Explorer, although the most used web browser, is not the only one. Firefox is now a well used browser. In October 2008 Browser usage was IE7 26.9%, IE6 20.2%, Chrome 3.0%, Firefox 44.0%, Mozilla 0.4%, Safari

2.8%, Opera 2.2%. Available at: http://www.w3schools.com/browsers/browsers_stats.asp [Accessed 12 November 2008]

13 Keep forms to a minimum and get rid of unnecessary questions. Make sure the form supports autofill.

14 Try to have a postal address on the site, as it adds credibility and security for users.

15 Don't use fixed widths on a site especially if the site will not fit on smaller resolution monitors.

16 If you are selling items on a website, offer a facility to view enlarged photos and when you show them an enlarged image make sure it really is enlarged and not just 10% or so bigger.

17 Make sure your 'Error 404: Page not found' will guide users back to your site.

18 A breadcrumb trail ensures a user knows where they are in a site. A breadcrumb trail goes across the top of the page's content and looks like this Home > Usability > Breadcrumb Trails.

19 Give a user suggested links on where to go next on a page.

20 Make sections on your website and define them clearly.

There are many more such standards that improve the site's usability but it is important to embrace these standards. Trying something new is like putting the steering wheel for a car in front of the passenger seat and the ignition above the rear mirror.

There are advanced ways of checking site usability, for example webcredible.co.uk have an analysis suite where they can apply eye tracking technology. Eye tracking sees where the eye moves as it looks over a web page. You can have anybody review your site and the system will create a heat map or path showing you where the eyes were focused on a page. If this is combined with user feedback it can go a long way to showing you how users use your site. Eye tracking technology allows us to see how people interact with pages. Position, shape, size, colour, luminance and many other factors on a page will dictate how our attention is drawn to areas of a page.

GMN viewpoint

Building a site is one thing but making it work for you is another. If people find your site hard to navigate, or hard to achieve what they have set out to do, they will quickly move to another site. This will mean that all the money you put into having the site built and driving traffic to it will be wasted. The old saying you can lead a horse to water but you cannot make him drink seems to apply quite nicely here.

6 How to get a website built

I have set up a number of points below that should be taken into consideration when choosing a design agency

1 Ask colleagues and friends for recommendations.

2 Remember the person who comes to see you is a salesman. It's easy to over promise when selling a website. It's one thing for a salesman to say something is possible and another for a developer to actually do it well.

3 Look at a few options, as there are a lot of pre-built systems out there. See if you can find one that does the job you require or can be modified to do so. It is easier than starting from scratch.

4 Ask the agency to show you anything they have done before that is similar to your requirement.

5 Ask for references – so many people forget to do this!

6   Consider if the agency are going to be able to service your future requirements, updates, content management, design changes. What happens if that agency gets busy? Ask how they will maintain a high level of service. Most agencies will always have a capacity issue. Good developers are hard to find and it is important for an agency to keep them busy. You'll expect to pay a premium if you want very high levels of ongoing service. Consider how it would affect your business if you requested an amendment to the site that took a few days or a few weeks if the update is more complex. Most of the time it does not cause a major impact and the premium cost is not worth paying.

7   Does the sales person have any technical ability, if not how can they really understand your requirement?

8   Ask to see what process an agency has for developing websites. If they don't have one, run away fast. Commissioning a website is perhaps one of the most complex projects a company can undertake. Large web projects require careful planning, execution and support. Here is a real world example:

(a)   You want a website, one that provides a classified system that allows users to sell second hand cars. You want it for a couple of thousand pounds. The agency say no problem and get busy building your site (does this sound familiar?). They haven't asked you any more regarding your requirements and a week later they show you a website that lets someone enter an item on a website, listing its title, description, price and a contact number. You see this and ask where is the image of the car, and wouldn't it be good if you could have more than one image. You also say you want people to be able to search for a car by make, model and year. The list goes on. The agency should really have clarified this with you before the build was started. Either way you have a problem now. The agency have developed what they have quoted for so either this becomes an argument or the agency charge you more. Lead times go out of the window and the developers now need to butcher the code they have written to try and twist the system to meet your requirements. If the agency simply agree to your requests and carry on making changes they have again missed the point! They need to clarify and get in writing exactly what you want. They should go back to square one and find out everything you want the site to do before letting a developer put another finger on their keyboard.

(b)   Agencies should have a process something like this. First is the production of a Top Level Specification, outlining the general components required to create your website. From this they should produce a complete wireframe (a page by page overview of the proposed site in a simple visual presentation) along with a technical specification. This creates the blueprint for building your website. It removes any ambiguity as to what is going to be built. If it is not in the blueprint then expect to billed for any changes requested so make sure you really, really, really understand what the agency is showing you. The agency is there to build **your** website but only **you** will know exactly what you want built. The wireframe process will need input from all the areas of a company that might be affected by an online presence. This often crosses all departments of your business, each with their own requirements, so get them to check the wireframe too. Once the implementation of the site is completed a site will require a sign off against the original wireframe specification and any relevant change documentation. Content should be checked and proofed, so clarify who is going to be responsible for this, the agency or the client. Professional proofing services are available but will have a cost implication. Most agencies recommend a soft launch which puts the site into a real world testing environment. Expect there to be some bugs, but so long as they are rectified quickly and without fuss, bugs are a normal part of development. Shouting at developers is usually counter productive; their minds work differently from everyday users. Websites are complex systems and things can go wrong, so be prepared.

(c)   Even though the agency has a process for building your site you should have one of your own for testing and checking it. So many clients think that the agency and developers will make sure a site is perfect, but remember we are all human and the ultimate responsibility is yours. Don't leave it to anyone else to be sure your site is up to your expectations.

9   Do your agency understand the principles of web marketing? Once they have built the site will they be able to drive traffic to it for you? Search engine optimisation involves technical implementation at the core of the development stage to be successful.

10   What support packages does the agency offer?

11   What training does the agency offer once the site is built and does it cost extra?

12   Who is providing content on the site? Are you expected to write the copy and provide the images?

13   Once the agency has designed and built the website, who owns the source code? The answer to this will usually be that the agency owns the intellectual property (IP) of the code, not you. You will usually own a permanent licence to use the software created. Some agencies will allow you to buy the IP for a price. Make sure you are clear on this arrangement at the very start.

14   Who hosts your website, what do you get with that hosting and how much does it cost? Cheap or free hosting will usually not include any sort of disaster recovery solution. Is your data backed up, can you recover previous files and what happens if the server you are hosted on fails?

15   Does the agency outsource the development or design? If so to whom and have they worked together before? What examples of sites can you see that have been produced by this team?

6.1 In house development

There are a number of web building programs out there. The commercial application for building websites is Adobe's Dreamweaver. It is a complex piece of software but if you are building a simple HTML (Hypertext Markup Language) website, Dreamweaver has some presets that helps. Core developers tend to use Visual Studio from Microsoft.

If your business has the resources to hire developers then there are a few things to keep in mind. Developers are a high paid resource, setting you back anywhere from £30k - £100k per year each. They also need full time management because if projects are not managed correctly they can become confused and delays often occur, causing greater cost and time impact. Having in house support is a good thing but usually it's best to leave bulk technical development to established development teams. If you do decide to go down this road then remember the processes mentioned previously, as they will help in maintaining control of your project.

GMN viewpoint

If an agency cannot show you an example of a site they have built that might be similar to your requirement then be prepared for the development to take a little longer. A good agency should be able to build anything, but if they have not done it before they are bound to find hurdles that slow them down. If the price sounds too cheap, it probably is! Either you are going to have to pay more later or the agency will become demotivated, as they will make a loss on the job. Developers are not designers and vice versa. If an agency is full of web developers don't expect the graphics to look great! If an agency says they can do your Search Engine Optimisation ask them to show you 5-10 sites on page one of Google. Or a good number of search results for one site if they have not got 5-10 sites.

6.2 Hosting

Hosting is when a website is held on a computer (usually a server) that is connected to the Internet. Your website is simply a collection of computer files, the server is setup in a way that allows people entering your web address (URL, or Uniform Resource Locator) to view your site.

Domains: You or your agency register a domain name, domain.com for example. Then this domain can be pointed to a server by allocating the domain the destination server's IP address. The IP address (Internet Protocol address), is the unique address of each individual server on the Internet and generally looks something like this 195.122.50.79.

It is recommended that a business registers its own names; this way they will always retain control over their web addresses. Pointing a domain to your website will involve downtime if a website is already running, usually about four hours. Transferring a domain TAG can take 24 - 48 hours, a serious inconvenience if you need your email over this period of time.

Many people believe that it is necessary to host your website on a UK server or you will damage any chances you have of search engine optimisation. Google states that if you have a .uk domain, Google will know not to exclude you on a UK search. If you have a .com address hosted on a UK server, again Google will not exclude you on a UK search. Finally, if you have a .com or .net domain and you are hosted outside the UK then you simply need to tell Google which country your business (site) is based. This is done through webmaster tools. http://www.google.co.uk/webmaster (YouTube.com, 2008)

Web hosting solutions range from DIY hosting of web space from £2 per month to dedicated and managed solutions costing thousands. There is a big difference in what you get but it is important to know what that difference is. Take into account how important your website is and what is the consequence of your website going offline.

Work on a 'Downtime Scale' of 1 – 5

1   Nobody will worry, the website is used internally or referred to by word of mouth

2   It will be a pain but we can live with some downtime

3   Clients will not be impressed but it will have no serious impact on our business

4   There is a serious concern as it will reflect badly on our business

5   It will have a serious impact on our business and could result in a loss of revenue and company reputation

Remember the Internet is an inherently unstable environment, so avoiding all downtime is quite an undertaking. Servers are, after all, computers and there are many other factors that can interfere with performance and presence.

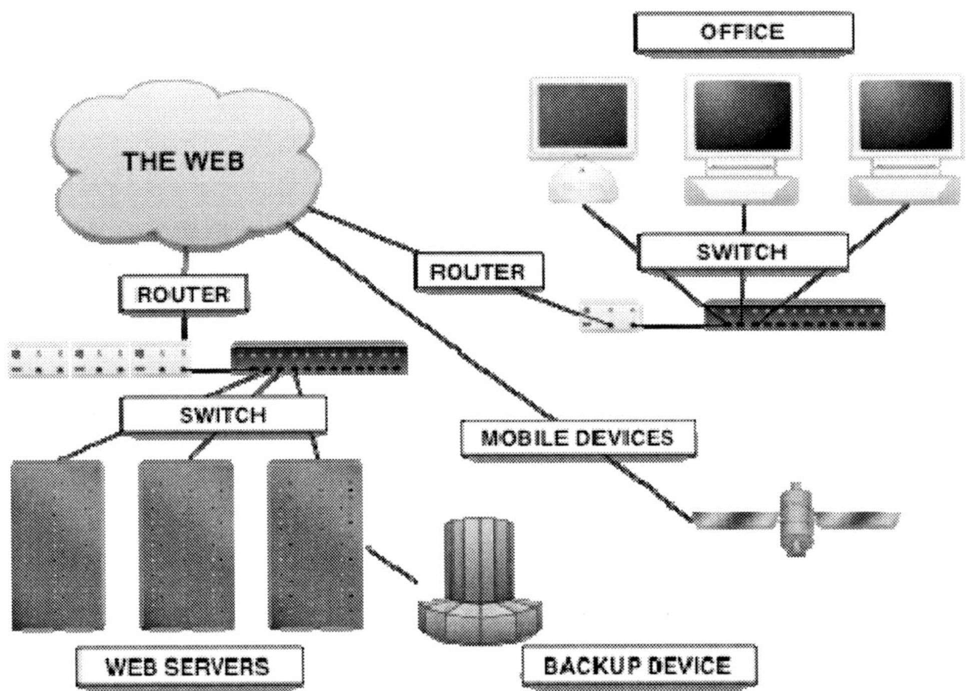

Typical diagram of fail points when hosting a site

1   A server can be subject to viruses, make sure your provider has anti-virus software.

2   A server has a power supply which runs 24/7 but they sometimes break. Make sure your provider has two supplies into their machine and a UPS (uninterrupted power supply).

3   The data feed into your ISP can get interrupted. Do they have a backup?

4   The bandwidth at your ISP many become over subscribed, causing your website to perform badly or not at all.

5   ISPs can fall under a denial of service attack or similar malicious action.

6   On a shared server new developments could be implemented that cause the server to crash.

7   Routers and switches are used to connect your server to the main data feed. These pieces of hardware can also fail.

8   Server hard drives can fail. What kind of backup strategy does your provider employ?

9   Servers can simply have a serious meltdown (not a very technical term, but basically something like a failed mother board). What disaster recovery plan does your provider have and how long would it take to implement?

10   Human error, someone trips over a plug, presses the wrong button etc.

So the real question becomes what are you paying for with web hosting?

The very cheap hosting is usually as a loss leader trying to attract customers in order to sell additional services afterwards. Cheap offerings are often cheap for a reason. A few realities that cause hosting prices to be higher are listed below.

One of the most important factors of web hosting is Hosting Support. What happens if your site does go down, who can you talk to, how hard will it be to get hold of them and what will they do once they get the call? In order to provide good support it is necessary to retain good support staff. This has significance cost implications for the hosting company which have to be passed on to customers.

A server is held at a data centre. These hosting facilities vary from the server room in a small office to a five star luxury hosting facility on the backbone of the Internet. The better the data centre, the better their connection to the Internet and the better their redundancy systems are. As a consequence it costs more to host a server at a five star data centre and these costs again are passed on.

Cheaper servers running single processors cost less than the big beasts that some hosts use. This will affect the performance of your website. So again if a good host has good machines, then this is reflected in the cost.

Sell cheap and pile it high! Servers can only cope with so much traffic. Web hosts selling server space cheaply often put more sites on to one server. Again this will affect server performance and ultimately the performance of your site. The fewer sites on a server the fewer customers there are to share the infrastructure costs of that server. But fewer sites means better performance.

What services are available within your hosting? A simple Linux server hosting MySQL has minimal licence costs compared to Windows servers running a full version of SQL, which is always going to cost more.

Back-up solutions again bolster the cost. Things can go wrong with your site so how important is it that the data is up to date when you put your last copy of the site back on the server? Does your hosting plan have a back-up plan and if so what is it: Daily? Hourly? Real time? Can you roll back versions? Roll backs are great when you accidentally delete that big order or decide to hit the erase button on a section of the site not needed then realise you need it back a month later. Good practice is to have 30 - 90 days of data stored.

Disaster recovery: what happens if the data centre burns down or the host server has one of those melt downs? What plan has your host got to get you up and running again and how fast can it be implemented? Any plan under 24 hours is going to have serious cost implications. Ultimately, unless a new machine can be put on the same IP and installed with all the sites, the only other option is to set up a new machine at another location and have a domain name re-pointed. This will usually take four hours, although to transfer the TAG will take longer. More sophisticated solutions are available, like having multiple servers in multiple locations all mirrored in real time, and load balancing, which is where the system shares traffic between the servers. If a server goes down the others take up the load.

So if you want a simple, no frills site hosted on a disk in some unknown location you can pay £2 per month. But if you require a custom built site with good performance, support and control then you are looking at more. Just make sure the hosting solution meets the business's 'Downtime Scale' requirement. Just because the ISP (Internet Service Provider) guarantees 99.9% uptime doesn't mean the website will be up all that time – it just means the connection to it will be there.

7   Digital strategies

Having your digital strategy clear is the key to the success of your online presence. Do you want to be the country's biggest online wine retailer, or the first stop website to what is on in London? Perhaps even the most popular website for information on the performing arts. Any of these would be admirable choices. There are seven steps to defining and delivering your digital strategy. I call this process GETREAL.

GETREAL

Goal Setting
Every Detail
Time Frame
Responsibility
Engagement
Action Plan
Leadership

7.1 Goal setting

Goals are simply visions for which you write a finish date on a calendar. Without setting a goal, forming a digital strategy is impossible. So first of all it is necessary to look at a business and work out how best an online presence can enhance a customer or user's experience of your company or brand.

Global case study

Within the next section we will be refering to an extended case example which we call Good Ol' Toy Company'. Use this to help you understand the process and concepts. Try to imagine similar large toy retailers in your own country.

First of all define what the website should be focussed on achieving. Using the form (fig. 1) score each category with an importance from zero to 10. In our example we have looked at a mid-sized toy manufacturer, 'Good Ol' Toy Company', who make wooden pre-school toys that are then sold through major outlets such as ASDA and Argos.

Score category by importance 0-10 then total scores

Target	Categories					
	Marketing	Operations	Support	Public relations	Sales	Management
B2B	5	1	2	7	10	2
B2C	5	1	4	7	10	2
Internal	4	2	2	0	0	6
Associate	4	2	2	7	5	2
Total	18	6	10	21	25	12

Figure 1

As you can see the primary focus for the website is driving sales, public relations and marketing. So the next question is what strategy should be employed on the website to make help best achieve these results.

We use the form (fig. 2) as per the example below.

List activities in order of priority 1-4, you do not have to choose all activities

Category	Activities							
Marketing	Information	2	Data Collection	3	Promotion	1	Research	4
Operations	Process	2	Account Management	na	Quality Control	na	Customer Services	1
Support	Personal	na	Technical	3	Product	1	Information	2
Public Relations	Brand Management	1	Brand Awareness	2	Internal	na	Stakeholder	3
Sales	Retail	2	Promotion	1	Consultative	3	Service	
Management	Reporting	1	Customer	3	Flexibility	1	Revenue	na

Figure 2

To make sense of these forms we extract the data into form (fig. 3) as per the example below. Order the categories into primary and secondary using form (fig. 1) then assign activities in order of priority using form (fig. 2). Once this is done assign the primary focus target for that category using form (fig. 1).

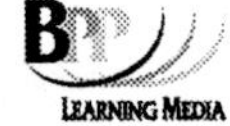

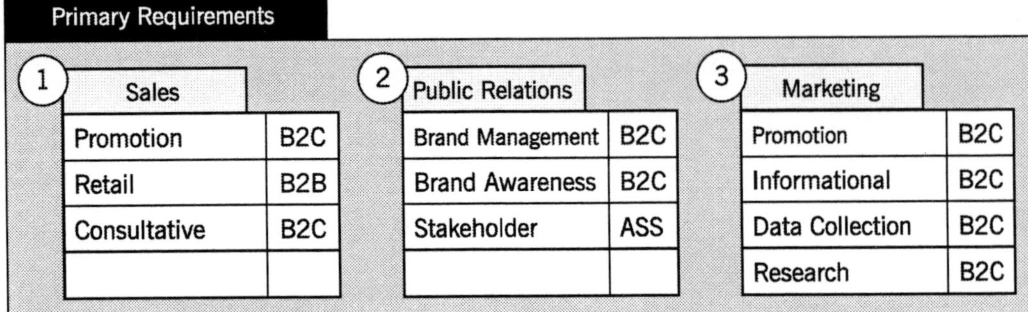

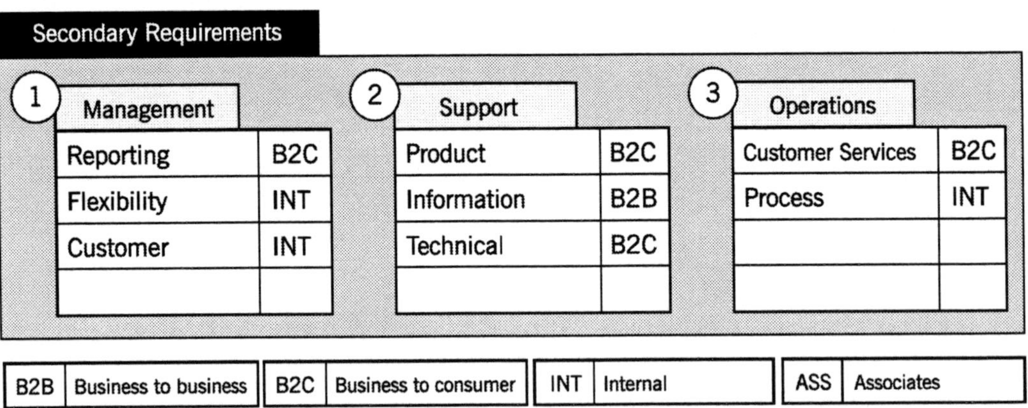

In order of highest score categories then list activities in priority

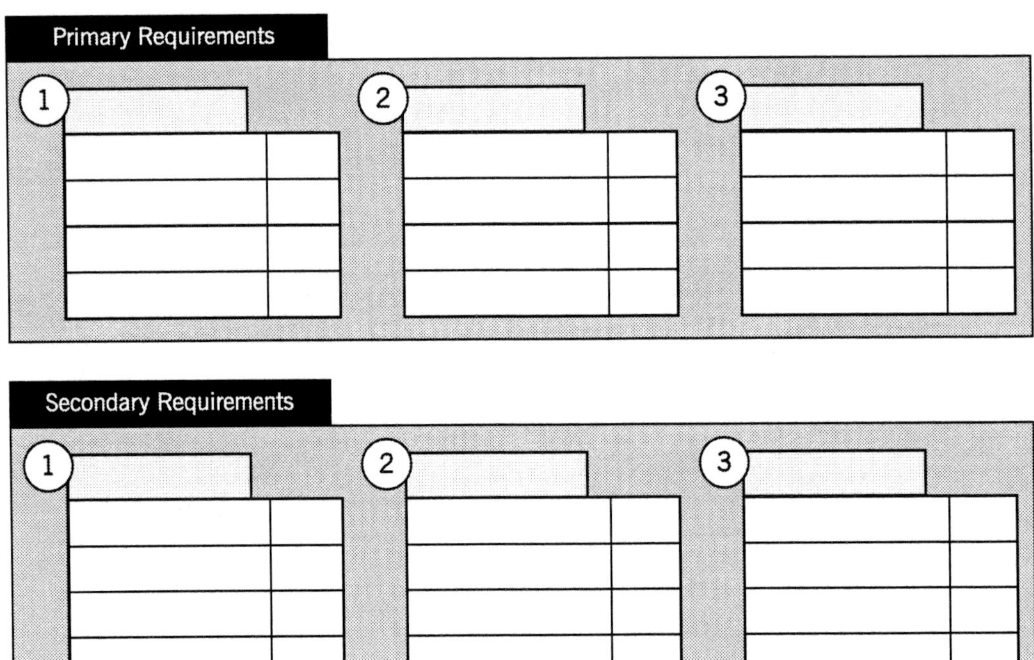

Figure 3

Example explanation:

Primary 1: Sales

The site's number one primary focus is to promote products to consumers, in order to increase value to the company's clients (the retailers); the site can be used to link directly to a retail opportunity with Argos for example. The consultative aspect of the site is to provide enough information so consumers can make an informed decision in their buying process.

Primary 2: Public relations

The site is to be used in managing the perception of the brand by current and potential clients. By creating a site that increases the business's credibility and shows a professional approach to managing and engaging consumers, you will increase brand desirability and ultimately brand equity. Using engaging content and defining a marketing campaign to drive traffic to the site will raise brand awareness. Providing a news/press section will help keep stakeholders up-to-date with the company's activities. Listed companies are obligated to publish stakeholder information on an Investor Relations section of the site.

Primary 3: Marketing

Advertising can be used to drive consumers to the site for additional promotions, in this case it could be a chance to win something. This process is ideally used in conjunction with a data collection process which can be used to build a future marketing list for the company. The data collection can have various questions which can be used to collect demographic and geographic information on consumers. Research into consumers can be best facilitated by ensuring a good analytics tool is tied to the website. I find Google Analytics (www.google.co.uk/analytics) by far the best. However, tools such as WebTrends (www.webtrends.com) and Weborama (www.weborama.com) also have some value as analytic tools. By analysing traffic data you can discover a lot about your site's traffic.

Secondary requirements should be considered if the primary focus of the site can be fulfilled properly using the company's current resources, leaving resources able to manage design, implementation and costs of these additional requirements.

'Good Ol' Toy Company' Goal Statement

"To increase brand loyalty and engagement resulting in a 25% increase in sales and development of a 5,000-strong mailing list by end of 2010" would be a suitable goal for 'Good Ol' Toy Company'.

Some other examples of goal statements

1   To become the number one online retailer of your chosen product with 50 plus sales per day

2   To become the number one information resource on a chosen subject with 100,000 visits per month plus

3   To cut down internal administration costs and time by 50%

Throughout recorded history, the most successful men and women have been those who've learnt to develop their natural goal-setting ability into a powerful skill for achievement.

I highly recommend Brian Mayne's Goal Mapping programme, which has been developed to powerfully impact your goals upon your subconscious. The programme is a combination of ancient wisdom and modern accelerated learning techniques, woven together with success principles into one holistic system which applies to everyone.

Mayne B., *Goal Mapping: an overview,* Lift International. Available at: http://www.liftinternational.com/goal_mapping/overview.html [Accessed 23 November 2008]

7.2 Every detail

Competitor analysis

Assessment advice

Before taking the next step it is necessary to research your competitors. In the case of the 'Good Ol' Toy Company' it would be good to investigate other businesses that match the company's profile. Below is a list of just 10 checks that can be performed to get a better picture of what and who your competitors are.

1. Take a good look around known competitors' online presence.

2. Once you have decided on the competitor who has the best presence on the Internet, carry out an alexa check on the site. www.alexa.com/site/ds/top_500 provides a popularity score for a website. Websites with a score under 100,000 are amongst the most popular sites on the Internet. Traffic rankings will tell you how popular the site is and related links will show you sites that are visited by the same people who view your competitor's site. This is a good way to reveal other potential competitors.

3. Check who links to your competitor's site. This can be done several ways but again alexa can reveal this information or you can do a simple search in Google using link:www.domain.com as a prefix. You may discover some sites that you would like to link in to. Links form a serious part of search engine optimisation.

4. http://www.compete.com can reveal traffic information to certain sites and

5. http://pr.blogflux.com/index2.php can be used to check a site's page rank. This is to show how important Google finds the website pages on a scale of 0 lowest to 10 highest.

6. www.who.is will give you registration details of a site including ownership and technical contacts

7. http://www.marketleap.com will give you a Link Popularity Check to show how many total links to your site there are across multiple search engines. Search Engine Saturation shows how many pages are indexed in main search engines and Keyword Verification lets you see if a keyword is featured in the search engines.

8. http://www.google.co.uk/blogsearch?hl=en will allow you to see what is being said about your competitor online

9. http://www.google.co.uk/news?ned=uk will show if your competitor appears in any news searches

10. It is also worth searching social media sites to see if your competitor has a presence eg Facebook, MySpace, YouTube etc

7.3 Choose your Revenue Model

7.3.1 Advertising

There are two main models for paid advertising.

1. **CPM, which is cost per '000 impressions**

    CPM, impression-based banner advertising costs vary across sites depending on the target demographic of the site audience and the value that has to an advertiser. A common problem with selling advertising on a per '000 impression basis is that often a website is unable to fulfill the number of page impressions to serve that advertiser's campaign within a set time period. Costs currently run from about £5 per '000 to £60 per '000. This is a very effective method for sites with large traffic.

2. **Fixed price and time period**

    Fixed cost advertising is used to allow advertisers to have a presence on a page for a fixed period of time. This can vary from a micro button for a year at £1,000 to a skyscraper banner for a month at £295. Cost should be calculated to ensure advertisers can see a good return on their investment (ROI) otherwise it is unlikely they will repeat advertise. For example, if you charge £2,600 for a presence on a whole page of your website for a month and the client is selling Plasma TVs with a profit of £200 they would need to sell 13 TVs just to break even. If the advertiser was selling cars the ROI would be easier to realise.

Some less common advertising models for websites are Cost per Click (CPC), Cost per Lead (CPL) and Cost per Sale (CPS).

CPC, CPL and CPS are often adopted by affiliate advertising. Affiliate campaigns allow you to select from a large inventory of adverts and display them on your site. Google Adsense is one network that will automatically display relevant adverts on your page which when clicked will generate a small amount of revenue.

See **www.google.com/adsense/**. Tradedoubler is a large network that provides advertisers with a distribution method to show their adverts. These can often be more rewarding in terms of revenue. **www.tradedoubler.com**. Finally some companies run their own affiliate program, for example **www.123-reg.co.uk/affiliate.shtml**.

Adverts usually take the appearance of banners; sizes vary greatly and you can see a list on **http://www.designerlondon.co.uk/page/Standard-Website-Banner-Sizes**.

7.3.2 Direct sales (e-commerce)

Selling products directly from your site is a great method of generating revenue. If you are a manufacturer, wholesaler or retailer this gives you a direct route to market. There are three main routes for selling direct online:

1. **Direct to consumer**. You supply products direct to customers, usually from stock. This means you have the most control over meeting clients' expectations regarding delivery. Key drivers for online purchases are price, delivery cost and delivery time. By holding stock and controlling this process, you usually have a greater chance of meeting customer expectations and growing brand loyalty

2. **In-direct to consumer**. Orders placed on your site go to third party suppliers to be fulfilled. This means that the stock liability is with the third party. It is a cost-effective method of running an e-commerce site and often means you can have much larger range options with less risk. On the down side, you do not have control of the third party stock levels or delivery timing. Also creating a

standardised branded packaging is more difficult as you will need to ensure that your suppliers deliver under your brand and not theirs.

3. **Digital Content.** Possibly the finest method of e-commerce is to sell a digital download in some form. This could be software, an e-book or audio track or classified or job posting for example. The download is typically delivered immediately after payment.

7.3.3 Lead generation

For service oriented businesses it is common to operate on a lead generation model. This utilises an enquiry form or forms. It is important when using this method to ensure you have the ability to track conversions on the forms. Also, critically, a way of tracking telephone enquiries from the web. There is technology that can be used to change a telephone number across a site depending on the source eg if the click comes from Google, for instance. At the very least use a different number on the website than on other sales material.

7.3.4 Membership and subscriptions

The holy grail of online revenue but possibly one of the hardest for directories, social media and premium content sites. If you have something that will make people want to subscribe to your site, then revenues can be very high. Eg a directory of just a thousand people paying £29 per month will generate nearly £350,000 per year. Sounds easy but usually the marketing costs to get these members is quite a high one.

7.4 Department involvement

Departmental input is critical in the early stages. When developing a website and online strategy it is important to realise it could affect many parts of the business. So it is important now to take this plan to each department and ask how each can contribute to achieving this plan. This process will allow a task and action list to be produced as a result. Getting the details right in developing an online presence and promoting it is the key to its success.

In the case of the site's development, be sure to have a rigid plan as to what is to be built. This will help manage expectations and give the site's development team a clear image of what is to be built.

Knowing each department's requirements will help ensure that your plan is thorough. It will be harder later to introduce additional functionality to a website or to request additional budgets for promotion of something unexpected. Below is an example of the departmental task analysis for 'Good Ol' Toy Company'

3.   Ensure they take ownership of their role in achieving the strategic goal. Let them own it!

4.   Most important, make sure they believe it, allow them to see the complete plan and how each part carried out will take the company one step closer to that goal.

7.9 Action plan

Now you have your task list it is necessary to work out your priority for each task. This should be looked at in conjunction with your time frame. I use a scale of 1-10 as there are often many tasks involved in the complete fulfilment of a digital strategy.

This can be done on the task/action form example below.

Task/Action	Priority	Who	Digital tactic/solution
Get data from commerce department for product availability	2	- J. Smith - Dev Team 1	Get data and build Locate store system
Get data from commerce department for potential and existing retailers and distributers	2	- J. Smith - K. Faulkner	Get data and run Mass email campaign
Define products for online competition based on retailers best selling products	4	- E. Woods - Dev Team 1	Specify competition and system

Once this has been done it is time to group all tasks by priority and schedule each item either on a calendar or project plan. It may be good to schedule email reminders to be automatically sent to each team/person as their deadline approaches. There are many tools ranging from Outlook to services online like www.memotome.com or www.poingo.com.

7.10 Leadership

It is the ultimate responsibility of the marketeer to champion the strategy. This should involve regular meetings and follow ups. If outsourcing to a third party, expect to have to make a fair amount of time to monitor and analyse activities. There is a big difference between delegation and abdication.

8   Digital tactics

8.1 Introduction

Once you have a strategy, it becomes necessary to define which tactics you wish to employ to realise that strategy.

Below is common guidance on the ten main factors that quantify individual marketing tactics. They are for guidance only as all services can be obtained at either end of the scale. Viral marketing, for example, can cost as little as the design of an email voucher or as much as the high price of the creation of an exciting interactive website such as http://www.elfyourself.com.

Definitions

Targeted: The target audience is specific to your service, information or product.

Flexibility: To manage change, targeting and cost of a digital tactic.

Cost: The potential cost of the digital tactic in relation to alternative marketing methods.

Conversion Cost: The typical level of conversion cost per enquiry/sale based on a digital tactic. This is affected heavily by the type of product or service being sold. For example, if you are selling high-margin wine and offer a viral of a 40% off voucher you will find your conversion costs will be very good, especially compared to a "10% off your next print" job voucher.

Brand Awareness: This is how well a digital marketing tactic will generate brand awareness.

Proactive Marketing: Many tactics are more reactive than proactive; for example, search engine marketing will only show a result when someone looks for a keyword.

Implementation Speed: How fast a tactic can be employed.

Long Term: How long the tactic will last once it has been implemented. PPC is an average long term strategy as it will only show your advertising for as long as you pay for the service. Unlike SEO, which can keep you in the search engines for many years if implemented correctly.

Measurability: The ease with which you can measure the results of the tactic.

Understandability: This is based on how easy it is to understand the tactic. Although SEO is the best long term strategy, it is also one of the most complex ones to understand, with many areas affecting rankings.

These guidelines are based on costs and results from a typical agency or service provider.

Scale	Bad	Not Very Good	Average		Good		Very Good
	SEO	PPC	SMO	AFFILIATE	BANNER	VIRAL	E-MAIL
Targeted	Good	Good	Very Good	Average	Good	Good	Very Good
Flexibility	Good	Very Good	Good	Good	Not Very Good	Not Very Good	Average
Cost	Average	Not Very Good	Good	Average	Not Very Good	Average	Good
Conversion Cost	Good	Good	Not Very Good	Good	Not Very Good	Average	Average
Brand Awareness	Good	Good	Very Good	Average	Good	Good	Average
Proactive Marketing	Not Very Good	Not Very Good	Good	Good	Good	Very Good	Very Good
Implementation Speed	Not Very Good	Very Good	Good	Average	Average	Good	Very Good
Long term	Very Good	Average	Good	Average	Not Very Good	Not Very Good	Not Very Good
Measurability	Very Good	Very Good	Good	Good	Average	Good	Good
Understandability	Not Very Good	Good	Not Very Good	Average	Good	Average	Good

8.3 PageRank (PR)

Although not a direct tactic, PageRank does play a part in many aspects of digital marketing. The PageRank algorithm was originally created at Stanford University by Larry Page, hence it is called PageRank.

Google Fun Facts. Available at: http://www.google.com/press/funfacts.html [Accessed 29th November 2008]

Google assigns to your web page an importance in the form of a numeric representation (from 0 to 10), with 10 being the highest value. The more sites that link to your site the more important Google considers your site to be, so this will help to increase your site's presence on the Internet. There are not many PR10 sites, here is a list of a few of them.

World Wide Web Consortium – http://www.w3.org

Google Search – http://www.google.com

National Science Foundation – http://www.nsf.gov

In simple terms, Google thinks that when one page links to another page, it is casting a "vote" for that other page. The more "votes" a page has, the more important the page. What makes it more interesting is that not all "voters" are equal. The importance of the page that's casting the vote determines how important the vote itself is. So the higher the PR of the page that's casting the vote, the more that vote is worth to your page. We must also take into consideration that the more links a page has going out, the less valuable those links become to the site to which they are linking. So if you have one link going from a PR 3 page to your website, that is a good link. If you had a link to your site from a PR 8 page, but there are 100 other links from this page, you would be sharing the link value of that page with 100 other sites – meaning the PR 3 page would actually be a better vote for your site.

PageRank is also affected by the internal linking structure of your website. This works very much like the external links calculation above. Each internal page has a PageRank that can be used to vote for other pages on your website. This is why contextual linking is so important; the more links to an internal page from other internal pages the better. You must also consider that outbound links from your page leak PageRank.

8.4 Search Engine Optimisation (SEO)

Although there are many ways of driving traffic to a website, the most cost-effective long term strategy is Search Engine Optimisation. SEO is not an overnight success so you can easily expect good SERPS (Search Engine Ranking Positions) to take six to twelve months to achieve. If done properly your presence in the search engines can be a long and happy one.

There is a common perception that 70% of all search clicks go through the organic (natural) listings, while the rest go through pay per click advertising. So having success via SEO can be extremely rewarding.

In this section we do not go into technical details about how to do optimisation, but I am going to tell you the 12 main factors that go into making a successful campaign.

First you should know that the top six leading search engines based on US Internet usage are www.google.com with 61.8% market share, Yahoo! with 20.6%, MSN 8.5% and AOL 4.5 and ASK with 4.5%. (comScore, 2008)

The UK picture is a little different with Google taking a massive 84.9%. (Hitwise, 2008)

Factor 1. Keywords

Picking the correct keywords is essential, but luckily there are some great keyword research tools on the web, like www.wordtracker.com or https://adwords.google.com/select/KeywordToolExternal. These tools let you analyse what words are commonly searched for, so even though you thought people were searching for "wall mount wine racks", a product you may happen to sell, you may not realise the search traffic for "hanging wine rack" is actually 10 times more searched! So if you optimised for "wall mount" your website won't be picking up the bulk of potential customers. So my first point is pick the right keywords, get them researched – don't just guess.

Factor 2. Inbound links

Inbound links are hyperlinks from another domain name to your website. They are counted by Google as votes for your website. Please do not confuse inbound links with reciprocal links (linking from your site to theirs and theirs to yours, these links will cancel out much of the link value). It doesn't stop there, as inbound links are better rated if they come from websites that have relevant content eg if I was a wine magazine linking to your wine rack website from a page that reviewed websites this would give a greater vote than me linking to your page from my branding agencies website.

Building links has to be done very carefully as building too many links at once will catch Google's eye and may have a negative effect. The same goes for adding inbound links from disreputable link farming sites.

To check in a search engine how many links there are to your site, type link: www.yourdomain.com.

Factor 3: Page Title

Page Title is the text that appears in the top bar of the browser window. This is also the line that appears as the blue link line in Google's search results.

Your primary keyword/phrase should be at the beginning of your page title to maximise on optimisation results. The title tag <title>Text Goes Here</title> appears at the beginning of the HTML code in

between the <head> </head> tags; a good Content Management system will allow you to add this information without having to access code.

Factor 4: Meta Tags

Meta Tags are code elements that provide structure to a page.

```
<TITLE>Online Trading Software System - Realtime Currencies, Indices, Commo
<META NAME="keywords" CONTENT="online trading, trading online, forex tradin
<META NAME="description" CONTENT="Download CCC's demo or live online tradin
<META NAME="rating" CONTENT="General">
<META NAME="language" CONTENT="english, EN">
<META NAME="charset" CONTENT="ISO-8859-1">
<META NAME="distribution" CONTENT="Global">
<META NAME="robots" CONTENT="INDEX,FOLLOW">
<META NAME="revisit-after" CONTENT="7 Days">
<META NAME="author" CONTENT="www.agi.co.uk">
<META NAME="publisher" CONTENT="www.agi.co.uk">
<META NAME="copyright" CONTENT="2006 Action Graphic International">
<meta http-equiv="Content-Type" content="text/html; charset=iso-8859-1">
```

Keywords: Although the keyword tag is no longer used by search engines, it is not conclusive whether having keywords makes any difference to optimisation. It is, however, still good practice to include about 20 keywords in this tag that also appear in the main content, title and description on the website.

Description: This tag does play a significant part in optimisation and again keywords/phrases should be placed at the start of the description. This information should also be written for the viewer as it appears as two lines in a search result and gives meaning to the page's content.

Other attributes such as language, charset and robots all act as guides for search engine robots (small programs that index your site). You can use a robot tag <META NAME="robots" CONTENT="INDEX, FOLLOW"> which tells the robot to index this page and follow any links to other pages while <META NAME="robots" CONTENT="NOINDEX, NOFOLLOW"> tells the robot not to index this page and not to follow any links off the page. As far as I understand, Google appreciates it when you tell it not to index certain pages as it means you have taken care in giving information to the search engine.

Factor 5: Heading Tags <H1> and <H2>

Heading tags range from H1 to H6 although the indexing of a site primarily takes notices of H1 and H2. It is common practice to create a page using a Headline (Content Title) and a Sub-Headline.

For example:

Heading Tag Search Engine Optimisation Guide

This Heading Tag guide will tell you the best practice for including H1 and H2 tags on a page.

Then you have your body text.

The code for this looks like:

```
<h1>Heading Tag Search Engine Optimisation Guide </h1>
<h2>This Heading Tag guide will tell you the best practice for including H1 and H2 tags on a
page</h2>
<p>Then you have your body text</p>
```

Notice how I have kept the keyword Heading Tag near the beginning of both sentences.

Factor 6: ALT Attributes

ALT attributes are text descriptions given to elements on a page such as images. Not only can they be used to add keywords to a page but they also allow people who cannot view images to see what the

image description is. They may not receive images because they are viewing on a device that does not show images, or they may have a disability and be using screen readers to tell them what is on a page. Search engines will use the descriptions to index the images on a page.

It is necessary to have ALT tags on images to comply with accessibility guidelines. The alt tag appears in an image <img src="images/homeicons/so.jpg" alt="Take me to Home Page" />

Factor 7: Optimised URL

Web addresses (URLs) should include keywords. If you were having a page on making birthday cakes, your web address would want to look something like http://www.birthday-cakes.co.uk/making_birthday_cakes.html

There is a technology called ISAPI (The Internet Server Application Programming Interface) which works with IIS (Internet Information Services) on the web server and can be used to rewrite a web address. This is especially powerful when used on a database-driven website with content management. It means that on a shopping site you might find a web address like:

http://search.barnesandnoble.com/booksearch/isbninquiry.asp?z=y&cds2Pid=17466&isbn=0545010225

Nowhere in the Barnes and Noble link do you have the title of the book, unlike Amazon below

http://www.amazon.com/Harry-Potter-Deathly-Hallows-Book/dp/0545010225/ref=pd_bbs_sr_1?ie=UTF8&s=books&qid=1203856597&sr=1-1

Wikipedia simplifies this url to just http://en.wikipedia.org/wiki/Harry_Potter_and_the_Deathly_Hallows

Factor 8: Sitemaps

Sitemaps provide a direct link to all pages on your website. When an indexing robot arrives at your site the standard way your site gets indexed is by the robot (or spider as it is sometimes called) skipping from link to link on your pages. After a while the robot stops skipping links and you end up with pages yet to be indexed.

In this example you can see how the arriving robot branches out across the site.

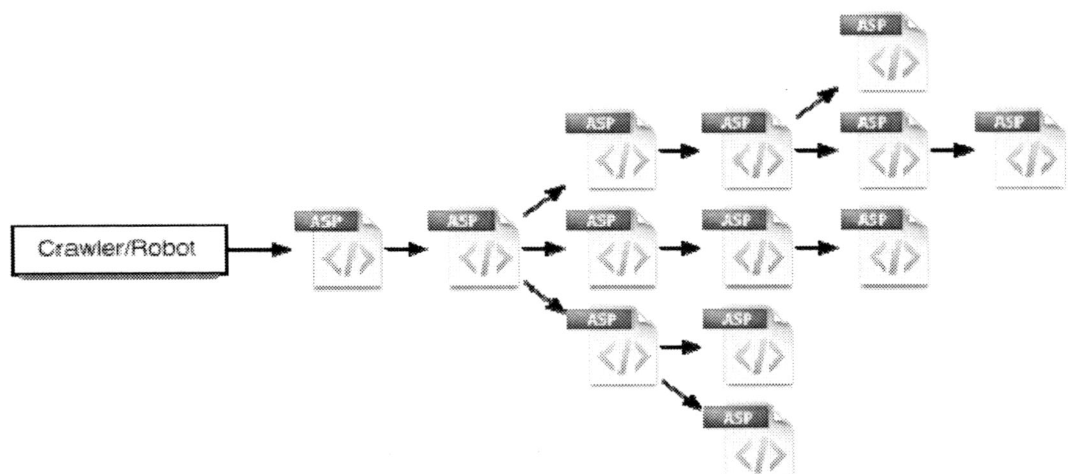

If you provide a sitemap.xml file in the root index of your site, the story is very different. Sitemaps were introduced in 2005 by Google as a way of organising information and allowing faster and more efficient indexing of a site.

In the example below you will see that the robot comes to your sitemap file and sees a direct link to every page on your site. The robot will index your website to a deeper level and in more detail if you have a sitemap. There is an online sitemap generator at http://www.xml-sitemaps.com/

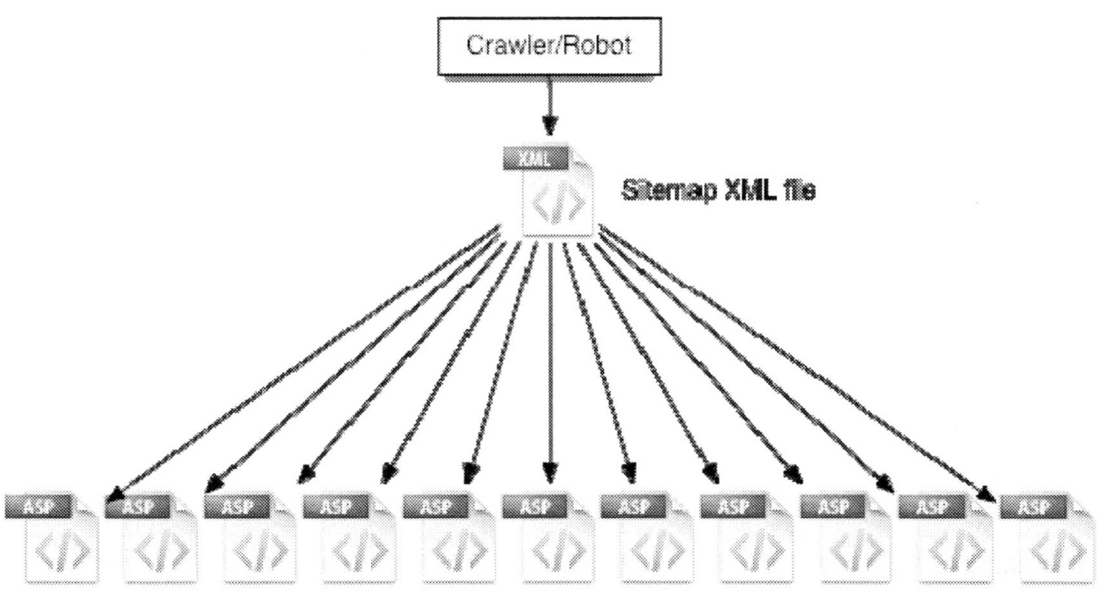

Factor 9: Robot.txt

The robot's exclusion standard is a way of telling a robot what parts of a website should not be indexed, however not all robots take notice of a robot.txt file. An example of a robot.txt file maybe

User-agent: *
Disallow: /cgi-bin/
Disallow: /images/
Disallow: /tmp/
Disallow: /private/

This would prevent the indexing of the directories as above. By disallowing the images directory the Google index would not include the graphic files from this site. So people who may be searching for images on Google would not have access to this site's images.

Factor 10: Keyword density

Keyword Density is the number of times you feature your keyword or key phrase in the text on your site. Depending on what keyword checker you use results will appear differently. A good checker is http://www.seochat.com/seo-tools/keyword-density/. In principle if you look at the code below you will find a keyword density of approximately 10%, which is ideally what Google looks for. It is very difficult to write copy that actually reads well and maintains a 10% density. The HTML example below is also repetitive so it would probably be considered Spamdexing by Google. I have used it purely to illustrate a point. If you have 200 words on a page and 20 of them are your keyword phrase, in this example "search optimisation and keyword density", then this phrase would have a keyword density of the magic 10%. Historical techniques such as keyword stuffing have a negative effect on your ranking, so don't try to be clever. Techniques like white text on a white background or hiding text underneath an image or adding in unrelated popular search words like 'sex' could get you banned from Google's index!

Some density analysers will count tags such as Titles and Meta so these items must include your keywords. This is an example of a visual density checker (www.iwebtools.com) analysing a website page for a serviced office company. As you can see the client's brand name is strong as a keyword, as well as geographical locations, but strongest are the words office, offices, serviced, location, guide, london.

0870 03122007 addresses all amp **avanta** avantas birmingh birmingham business can careers centre clients club completely concierge corporate do ec2 edinburgh facilities find flashtrade forward guide ha1 harrow heathrow help home i information leading locations london look macromediatrade mailing manchester market meeting members **midlands mtr** my need new north offering office offices opens our outsourcing overview pa packages particular place plugin prime private products quality reading requirements room rooms roomswhether run **scotland** service serviced services small solution solutions space street sw1 **thames** training ub3 uks unbranded unique us use **valley** virtual w1 w6 walkthrough wc1 wc2h we west you 24012008

GMN viewpoint

Write your text for the customer, make it interesting and useful.

The ideal amount of copy is 200 - 600 words.

If you reach 5% - 7% keyword density you are actually doing pretty well.

Try to get your keyword in the text once every 10 words.

Use your keywords at the beginning of your headline and your sub headline.

Factor 11: Validate your code

Although not necessary, making your page's code 100% perfect makes the page easier for a robot to crawl. Imagine reading a book and the writing is grammatically incorrect, there is no punctuation, and all in all it is not well written. Although you might get the idea of what the page is about, you won't enjoy reading it and you certainly won't understand it clearly.

Validation is done by W3C (org http://validator.w3.org/). The World Wide Web Consortium is there to develop web standards. It was set up and is run by Tim Berners-Lee, W3C Director and inventor of the World Wide Web.

At the time of writing, Amazon has over 1,000 code errors on its home page. So as you can see, although it is good practice even the biggest sites don't always comply with validation.

Factor 12: Contextual linking

Cross linking pages on your website using link words usually made from keywords or phrases, encourages greater indexing of the site and adds importance to those keywords. It is good SEO practice to have keywords in a piece of text that link to additional pages of content regarding those keywords. I have underlined some words in this paragraph which if on a website might be contextually linked to other pages.

Another way of adding contextual links to a page is to have footer links or *See Also* links, for example:

See Also:
Cross Linking Pages
What are Keywords
What is Indexing

There are many other factors that affect SEO, ranging from the age of your domain name to the location of your server. But the ones above are key.

GMN viewpoint

Get your keywords into your Title, Keyword and Description Tag. Get them into your URL, get them in your title (H1) on your page, repeat them in the content of your website and you will be a long way to getting on the first page of Google.

For some great information on how people search the web you can visit Zeitgeist, a little known site owned by Google that shows search patterns and trends – http://www.google.com/press/zeitgeist.html. The information is compiled from billions of searches.

8.5 PPC (Pay Per Click)

This is the fastest way to get your site a presence in Google, Yahoo or MSN. Facebook also do PPC, which means you can reach an amazingly targeted audience with your advertising.

Let's focus on Google Adwords, as Google makes up more than 80% of UK searches and over 60% globally.

The main focus for Google Adwords is the PPC advertising that appears on the top and right side of a Google page and is labelled sponsored listings. It's the main revenue generator for Google, who made over $16 billion in 2007.

Google Investor Relations http://investor.google.com/fin_data.html [Accessed 14th December 2008]

Google adwords also provides PPC advertising on Google Maps (Local Business Ads), Google Mobile, Image Ads and text adverts across the content network,

'The Google content network comprises hundreds of thousands of high-quality websites, news pages, and blogs that partner with Google to display targeted AdWords ads. Partners like about.com, lycos.com, thenewyorktimes.com. The Google content network reaches over 75% of unique internet users in more than 20 languages and over 100 countries. As a result, if you advertise on both the Google search network and the Google content network, you have the potential to reach three of every four unique internet users on Earth.'

Google Adwords https://adwords.google.com/select/afc.html [Accessed 14th December 2008]

Let's consider some of the basics.

Google adwords are set up using an Adwords account found at http://adwords.google.co.uk. The first thing to know is that just because you bid the highest price for your Google advert it doesn't mean you'll get to the top of the search engine – a position is based on an advert's quality score multiplied by the bid price.

What happens when you go beyond the basics?

To analyse the setup of accounts and keywords, use the Google keyword tool https://Adwords.google.com/select/KeywordToolExternal and the Adwords preview tool, which allow us to see how and where an advert will appear without it costing each click or counting as an impression. There are many more tools but the most important is the conversion tracking tool. With this we can actually measure the ROI of a campaign and quickly see the cost of acquiring each customer. The tool is simply a piece of code you place on a goal page eg a "thank you for your order" page.

Once you are using conversion tracking you can use the conversion optimiser, which will allow you to set a maximum cost you are willing to bid to acquire a customer. It uses historical information about your campaign and automatically sets your cost per click each time your advert appears.

Then it uses the Google crystal ball to work out which clicks are likely to be most valuable – the crystal ball in this case being a complex algorithm based on your conversion history.

This is great if you are running campaigns with lots of keywords and need to have each cost per click managed automatically.

Other things you can do with Google Adwords

Pay-per-action advertising: This is used on the Google content network, a large group of websites, email programs and blogs which have partnered with Google to display Adwords ads. You can bid to pay only if an action is fulfilled, be it a purchase, sign-up or completion of a desired process. You still compete to appear with cost-per-click and cost-per-thousand impression ads.

Cost-Per-Thousand advertising: This is known as CPM as M is the Roman numeral for 1,000. It's the means of fixing a price for an ad that appears on the content network per 1,000 times that advert appears. You will pay the same price for those 1,000 impressions whether you get 500 clicks or none. The bid price for the CPM will be factored into where your advert appears but Google will intelligently work out the minimum amount to charge in order to keep your ad's position on that page.

Getting your adverts to appear

If you are running adverts on both Google's search page and its content network, you can define a different bid rate for each area. So if you know you get more conversions through the content network you might want to raise the bid rate to increase your presence.

Adverts can also be shown in two ways: Standard delivery, which will make your advert appear at intervals throughout the day, and accelerated delivery, which will make your advert appear as often as possible until your budget runs out. So if you run a low budget you can get your advert to appear once every 1,000 searches throughout the day or every search until 10am, for example.

You can geographically target your adverts and even schedule the times for the delivery of your adverts to ensure your budget is not being spent at times when no one is looking for your advert.

What are you missing still?

Well, Google has many areas in which to advertise other than the standard search results. You have Google's search partners, AOL, Compuserve, Ask and AT&T to mention a few. There are the content network partners: About, Lycos, NYTimes, Reed Business etc. Google's Gmail is also part of the content network as well as many individual websites signed up to the Google Adsense program.

Adverts that can appear on these networks contain quite a variety that goes beyond the standard four-line text adverts. You can place banner adverts of all sizes, video adverts and mobile adverts which can include a click to call link.

You can also now place advert on Google maps so if you want to be found as the local Chinese takeaway or the closest tyre repair centre, Local Business Ads are great.

GMN viewpoint

Don't think you have to guess to set up a campaign. Google gives you all the tools you need to run successful campaigns, you just need to learn how to use them. Also remember it's not all about who's got the most money – although that can help!

As you can see, just sticking a single bid price on a bunch of keywords is not going to give you the best ROI. So if you are setting up a campaign yourself, take advice, learn about it and cover the main bases – otherwise get a professional to do it and make sure your investment pays off.

Four important points to consider:

1. Is the keyword relevant to the landing page?
2. Does the advert feature the keyword?
3. Have you thought of the negative keywords (words you don't want an advert to appear for)?
4. Is there a final page in a process to track the conversions?

8.6 Social Media Optimisation – reviewed

Also known as SMO, this is the latest way to promote you or your business. It takes into account social media, online communities and community websites. It covers areas such as RSS feeds, Really Simple Syndication, which is a method of publishing fast changing content for people to subscribe to. The ability to add your site to websites that allow people to discover and share content, such as Digg, Del.icio.us and Facebook, is a way of driving traffic to your site via large traffic sites that can share information which a network may find interesting.

8.6.1 Blogging

The publishing of regular entries on a subject, event, or personal commentary is another social optimisation tool, for example blogger and wordpress.

See my example http://damonsegal.blogspot.com

8.6.2 Micro-blogging

Twitter is a social networking and **micro-blogging** site that allows users to post very short (140 characters) updates or messages. It has over a million registered users, although research shows that only about 200,000 are active on a monthly basis (TechCrunch, 2008).

Top Tweets are mainly by individuals but there is potential for business, and here are ten ideas I came up with.

1. You run a nightclub and when Saturday night comes you find it's looking a little quiet. So you post a Tweet to your 'followers' (people who have subscribed to watch your Twitter entries) saying "free first drink to any entries in the next hour." This would go out instantly to phones, PDAs and PCs, alerting your social network to the offer.

2. You are a school or community-based organisation Twitter and you can quickly update parents about events and emergencies such as a closure or the late return of a school trip.

3. You subscribe to a music group or a professional speaker's Twitter and want to stay up-to-date with appearances, dates and times.

4. You subscribe to a network of IT professionals and regularly Tweet problems to your network for fast advice and solutions on how to solve problems.

5. You run a recruitment or contractor business and need fast responses for requests to fast placements. For example, you need five waiters tomorrow night for a last minute event or you need an emergency plumber to go to a call out.

6. A news-oriented business can post the latest news as it happens throughout the day or night.

7. You are working in collaboration on a global project and want to keep all parties up-to-date instantly with the project's status.

8. A major High Street retailer who wants to update customers with daily offers. Boots could publish a Tweet each time they are doing a 3 for 2 offer or ASDA could alert shoppers to in-store offers or links to special online shopping pages only available to Twitter followers.

9. A large corporate outsources a great deal of print and allows a group of printers to tender or bid for the work. These tenders can be placed on Twitter and then printers can give their best price to win the contracts.

10. Your professional consultant has a daily blog, but people need to visit your blog to read it. So you post the headlines on Twitter to alert your followers to the latest blog being posted.

8.6.3 Social networking sites

Sites such as www.linkedin.com, www.myspace.com, www.friendsreunited.com, www.plaxo.com, www.xing.com and www.facebook.com provide a way of building and promoting your business. I'll use Facebook for this example.

Facebook is a social networking website that launched in early 2004. It was founded by Mark Zuckerburg while he was a student at Harvard University. If you sign up to Facebook you can create a profile for yourself or your business and people can then become your friend, fan or supporter. This forms a social network which can then be used to interact with – whether to arrange Satuday night or promote your brand.

Pop groups, brands, people, films, restaurants, all have a presence on Facebook. If you look at the number one page on Facebook [Accessed in August 2008] it was "Barack Obama – US Senator for Illinois." He had 1,342,636 supporters he communicated with instantly for free!

People will feel they know him, it will seem to them on a one to one basis. Barack Obama also has a MySpace page with 451,179 friends. This strategy appears to have worked for him very well as he won the US election.

If you enter into this arena keeping your information up-to-date is very important. Too many people/companies start a page on a social network site then give up. The key here is don't leave it there, SHUT IT DOWN. It looks very unprofessional to have an out of date page.

Global case study

Places can also be promoted. The top place with over 377,000 fans is Disneyland and it does a really good job in engaging people to want to go there. Top Products are Apple Students and Victoria's Secret PINK, both with about half a million fans. Apple seem, as always, to have a slick approach to their page, using a strong marketing approach to create a community as well as slick adverts and promotion of products. Victoria's Secret PINK seem to use the page mainly as an advert with a link back to their site. Some promotional banners for products are also there but not animated like the Apple one.

Facebook www.facebook.com [Accessed in August 2008]

Facebook business opportunities

Facebook Platform: Applications abound on Facebook. I gave up when I got to 2,000! Social applications allow members to interact with their friends and businesses. Most applications I found were for fun but some have practical uses. The most popular application is Slide FunSpace (formerly FunWall) (20,964,181 monthly active users, by the time you read this could be 30 million!), which allows users to find and share videos, posters, graffiti and more. Second is Top Friends (17,075,103 monthly active users), which allows you to give and receive exclusive awards, show off your mood and keep tabs on the people you really care about.

Facebook Ads: Perhaps the best targeted advertising ever created, ComScore reports that Facebook attracted 132.1 million unique visitors in June 2008, compared to MySpace which attracted 117.6 million. Creating a Facebook advert is straight forward and you really can be very specific with your target audience.

If you were selling an alco-pop drink you couldn't go far wrong showing an advert to people between 18 and 30 years old in the United Kingdom who like Drinking, Clubbing, or Getting Drunk. Turns out at this time there are 340,420 of this demographic on Facebook! Given a suitable incentive you could get quite a few people trying your brand.

Facebook is working on other services which should make the website's marketing capability even stronger.

Facebook www.facebook.com [Accessed in August 2008]

8.6.4 Forums

These are online communities which share a common interest. Unlike blogs, a forum allows members to interact with each other on a message board. Messages are posted on the forum and other members add to the post (these are called threads). In order to manage these threads and prevent abuse of a forum, administrators assign people they trust as moderators. Forums are excellent for finding solutions to problems eg do a Google search for "how do I program my harmony remote" and you find http://forums.logitech.com/logitech/board/message?board.id=programming&thread.id=18173

Businesses run forums to help provide support for customers. It means customers can often find solutions to issues without having to contact a business support line. Many communities with similar interests use forums to help solve problems quickly eg Google search "Duplicate Meta Description Content" and you find http://forums.searchenginewatch.com/showthread.php?t=24559

Forums can be used to promote your products, services or events or you can start your own to improve a support service, for example.

8.6.5 Social book marking

Social bookmark sites are where people can promote and share content that they find interesting. For example http://digg.com/users/damonSEOsegal is a Digg profile. In this profile you will find articles on digital marketing and other subjects that I find interesting. Business and individuals can have profiles on social bookmark sites and place links to those profiles. Buy placing a link in a bookmark site it allows you to promote web pages that you want to be found. For example if you were launching a new product and knew you had a following on a social bookmark site, you could place a bookmark to your new product. On a site like Digg the more people who bookmark your site the higher your submission is placed. With http://delicious.com/ this is a straight forward bookmark tool, which simply adds a bookmark link in your list. You also have the ability to allow a network of people to subscribe to your bookmark list. http://technorati.com is another bookmark site but this one is dedicated to bookmarking blogs. It allows bookmarked blogs to be indexed searched by users.

Links from social bookmark sites also help as part of the link building process for SEO.

There are many of these bookmark sites and in principle they do the same thing. Services exist that will bookmark your website, blog or press release across these bookmark sites. It can form an effective part of a link building and SMO campaign.

The top 10 bookmark sites as of December 2008 are:

1. Digg
2. del.icio.us
3. Technorati
4. StumbleUpon
5. Reddit
6. Furl
7. BlinkList
8. Ma.gnolia
9. Rojo
10. Simpy

8.7 Online PR (public relations, press releases) and articles

Online PR is a great strategy for driving more qualified traffic to your website. The placement of press releases and articles helps to increase visibility and search engine rankings by forming high quality inbound links to your website.

There are many online press release hubs, the leader in online news and press release distribution being PRWeb.com, but there are many others such as free-press-release.com, prlog.org, marketwire.com and webwire.com to mention just a few.

The trick to a successful online press release is making the release newsworthy; although some newswire services are automated, most are picked up by real people. You need to make sure that you are providing the media with something interesting about your organisation, product or service.

Make sure you have a grabbing headline which should hook the reader by outlining the story. Try to keep headlines under 80 characters as this will be displayed as the title tag in a search engine.

The first paragraph should be the important information, so get across what it is you want to tell people and make it exciting. "XYZ have just partnered with ABC resulting in 123", "300% increase in sales due to fantastic new feature" or "Consumers hit website in record numbers because..." No one wants to hear about what you do or how wonderful you are, it just isn't news. Make it personal and try to relate it to a real life story.

The next part of the press release should give the facts; journalists are naturally sceptical so making wild claims and making your story sound too good to be true, even if it is true, is likely to get your article a fast track to the trash folder. Also never use exclamation marks unless they are part of a quote as this will also get you canned!!!

See what is topical in the news and find out how to make your article associate with it. Make your verbs and adjectives strong – delighted not happy, outraged not angry or miserable not unhappy – as this will help engage the reader.

Keep away from technical words and acronyms. Remember that not all readers will be up to speed on your industry jargon. Keep your article understandable.

Once you have written the article, go back and remove all the unnecessary words. Make it more concise and follow the write less, say more philosophy. The more concise the message, the more likely it is to be understood and remembered.

Press releases should be between 300 and 800 words; you should not include an email or web address in the body of a press release but save this for the last paragraph, where you can include details about the company or person who the press release is about. Finally don't forget your contact details.

GMN viewpoint

1. Make it newsworthy
2. Make the headline grabbing and under 80 characters
3. Make the first paragraph have your key information
4. Don't make it about you or your company, tell a real life story
5. Keep it factual
6. Don't get too technical
7. Keep the press release between 300 - 800 words
8. Keep it concise
9. Use engaging words
10. Remember your contact details at the end

8.8 Affiliate marketing

This type of marketing has been around since 1994, and is a way of getting groups of affiliates to feature a link from their site through to another site where a click or a sale resulting from a click means the linking website owner is paid a commission. The link normally takes the form of a banner and usually re-numeration is paid on a Cost per Sale (CPS), but occasionally it might be paid on a Cost per Action (CPA) or Cost per Click (CPC). CPC is not recommended as it is open to click fraud where affiliates click their own adverts to generate revenue. The setting up of an affiliate program is a low cost risk if commissions are paid on a success basis.

In-house Affiliate Programs: This is where a target site implements its own programme for allowing people to sign up to an affiliate system. This system will create an affiliate ID which is placed into a unique link that is used by that affiliate. The affiliate then usually has a selection of banners to choose from to place on their own site(s). The unique link allows the target site to track which affiliates' clicks turn into sales.

Out-source Affiliate Programs: There are many third party sources that can be used to implement an affiliate program, and companies like tradedoubler provide advanced tracking technology. http://www.tradedoubler.com and http://www.affiliatemarketing.co.uk are Europe's leading affiliate marketing businesses and Tradedoubler alone has over 100,000 publishers. It works by utilising a network of website publishers called affiliates. Affiliates are rewarded per click, per unique visitor, per registration, per sale or for any other activity you specify. Using this same network some affiliate programs can be orientated to track phone calls enabling a pay per call fee structure to the programme.

8.9 Banner advertising

Banner advertising is common across the web. Banners are placed on a web page either directly by the web site owner, an advertising network or via a banner exchange. Adverts come in a variety of formats and sizes, anything from a static jpg graphic to an expanding rich media flash banner. Video banners are also now becoming more popular as the click through rates (CTR) are higher than those of standard banners. Generally the feelings towards banner advertising are quite negative and after many years of having them thrust on us we have learned to literally turn a blind eye. We automatically filter out banner ads, which is why it is important to make them as engaging as possible.

One important factor to remember with banner advertising is that your banner is usually appearing on a specific site that is relational to your product. eg http://travel.msn.com would typically show banner ads to do with travel and holidays, http://www.decanter.com/ would have adverts related to wine. People who do click on these adverts are usually qualified leads. Conversions of the limited number of people who do click through is traditionally high especially if the banner leads directly to the related landing page.

Banner ads are usually sold based on CPM (Cost per thousand impressions), which means that each time your advert appears on a page one impression is deducted. Costs vary on CPM but typically range from £5 per '000 to £60 although costs can go either side of these figures. Ads can also be sold on a cost per action (CPA) or cost per lead (CPL); this is a great option if you can find it, although it does not work as well for the advertiser.

When buying advertising on a CPM basis, be careful not to over purchase on inventory. Many websites promise thousands of impressions in a timeframe but cannot guarantee that those inventories can be served. Ask to see the latest page impression statistics for a site if you have a doubt that the site will be able to provide the traffic required to fulfill an order. Many sites that have low traffic may sell Banner adverts at a fixed price; there is an argument that just because these sites do not attract enough traffic to make selling by CPM viable they may still have very relevant visitors for your product or service.

Standard sizes

Rectangular and pop-up ads	**width by height**
(in pixels)	
Large rectangle	336 by 280
Medium rectangle	300 by 250
Square pop-up	250 by 250
Vertical rectangle	240 by 400
Rectangle	180 by 150
Banner and button ads	
Leaderboard	728 by 90
Full banner / Impact banner	468 by 60
Half banner	234 by 60
Button 1	120 by 90
Button 2	120 by 60
Micro bar	88 by 31
Micro button	80 by 15
Vertical banner	120 by 240
Square button	125 by 125
"Skyscraper" ads	
Skyscraper	120 by 600
Wide skyscraper	160 by 600
Half-page	300 by 600

The Interactive Advertising Bureau (IAB). Available at:
http://www.iab.net/iab_products_and_industry_services/1421/1443/1452 [Accessed December 16 2008]

To work out a potential % return on your investment with regard to banner advertising use this formula

Total Monthly Impressions / 1000 = CPM Units
Cost per '000 Impressions * CPM Units = Total CPM Cost
Estimated CTR% * Total Impressions = No. Click Through
Estimated Conversion % * No. Click Through = No. Conversions
Profit Per Conversion * No. Conversions = Total Revenue
Total Profit - Total CPM Cost = Actual Profit
Actual Profit / Total CPM Cost = CPM advertising ROI%

eg

5,000 / 1,000 = 5
£100 * 5 = £500
1% * 5,000 = 50
10% * 50 = 5
£1,000 * 5 = £5,000
£5,000 - £500 = £4,500
£4,500 / £500 = 900%

8.10 Email marketing

Email marketing is the modern version of direct marketing and replaces all those envelopes that have been landing on your doormat. Most providers of demographic data have their own mass email systems and will send and report on mails direct for the client. You can also purchase software to send your own mass mail. Software like Outlook is not good for sending out bulk email. There are also a number of online services such as http://www.mailchimp.com.

When sending mass email it is best to take small quantities of your audience and test various subject lines and headlines for a mail. Also timings: what is the best day or best time of day to send your email? Perhaps direct mail to purchase outdoor furniture should be sent on Friday morning when the weather report looks good for the weekend. The subject line might be 'Mr Smith, dinner in the garden tomorrow' with a headline 'We guarantee to deliver your outdoor furniture before 12pm tomorrow so get that BBQ ready!' In order to be able to react with that kind of speed you might find it best to have your own online email tool. Remember that to avoid upsetting recipients you should include a simple unsubscribe process at the bottom of your emails.

To maximise your emails' open rates perhaps the biggest key is that subject line; where appropriate it should be personalised and motivating. Ensure the From line is recognisable or applicable. Make sure your target demographic is applicable and that your list is not out of date. Also to avoid falling foul of an email software's junk filter, avoid words such as "money back ", "cards accepted", "removal instructions", "extra income" or "free" (Microsoft, 2008).

According to research conducted by the Direct Marketing Association, email marketing generated an ROI of $48.34 for every dollar spent on it in 2007. The expected figure for 2008 is $45.06, and the prediction for 2009 is $43.52. As such, it outperforms all the other direct marketing channels examined, such as print catalogues.

Research conducted by Shop.org in 2007 revealed email generates sales at an average cost per order of under $7, comparing favourably to $71.89 (banner ads), $26.75 (paid search) and $17.47 (affiliate programs.) (Email Marketing Reports, 2008)

8.10.1 Electronic mail basic regulations

Electronic mail is emails, SMS (text), picture, video and answer-phone messages. Electronic mail marketing messages should not be sent to individuals without their permission unless all these following criteria are met:

1. The marketer has obtained your details through a sale or negotiations for a sale.

2. The messages are about similar products or services offered by the sender.

3. You were given an opportunity to refuse the marketing when your details were collected and, if you did not refuse, you were given a simple way to opt out in every future communication.

The regulations do not cover electronic mail marketing messages sent to businesses.
(Information Commissioner's Office, 2003)

Here are eight reasons to use email marketing

1. You can reach the desktop of a potential consumer immediately.

2. It leverages your efforts – email marketing simultaneously presents your best sales pitch to thousands of people.

3. It allows you to target with precision – unlike most advertising email marketing it lets you pinpoint your customers.

4. You get an immediate response – within 1-2 days of mailing 80% of your results will be known to you.

5. It's easy to track your return on investment by tracking your email opens and link clicks.

6. It's relatively inexpensive.

7. It gives you the best chance of attracting the personal attention of your prospective consumer.

8. It can lead a consumer directly to a website where they can make an immediate purchase or enquiry.

All the above reasons make direct mail a very powerful marketing medium which, if handled correctly, will give a very high return on your marketing spend.

8.11 Viral marketing

Commonly used for brand awareness, viral marketing in a digital sense takes the form of an email or a web address with compelling content eg video, flash game or special offer. The aim is to target people with a high social networking potential (SNP) and have them pass on the marketing message to friends and colleagues. Launching a viral on a social media tool is a great way to kick start a viral campaign. For example, you belong to the Pepsi Max fan group on Facebook and they launch a series of funny extreme sport videos all branded with Pepsi Max. If you found the video worth watching, the chances are you will recommend it to your network and they to theirs and so on.

Global case study

A noteworthy example of viral marketing is the Threshers 'family and friends' voucher for 40% off, which was first sent out on email in December 2006. The voucher was downloaded by millions of people. Although a spokesperson for Threshers said 'It could end up hitting our profit margins' Threshers did only discount off full retail price and not in conjunction with any other offer. The cost of implementing this viral would have been minimal.

BBC News *Web Discount Frenzy at Threshers.* Available at:
http://news.bbc.co.uk/1/hi/business/6198828.stm [Acessed December 28 2008]

ElfYourself www.elfyourself.com is a online viral from OfficeMax, a large American stationery supplier. ElfYourself was visited by 193 million people in 2007. In 2008 this viral site was revamped and now works as an additional revenue generator selling you the video it creates or puts your image on Elf products such as mugs, mousepads etc. More than 40% of the people who tried the online app gave OfficeMax credit for bringing goodwill. Thirty per cent of those said they would shop at the retailer. The implementation of a flash website like this would have been costly, to say the least, but the brand exposure is excellent for a multi-billion dollar business like OfficeMax.

MediaPost News *Many Happy Returns: OfficeMax Brings Back ElfYourself.* Available at:
http://www.mediapost.com/publications/?fa=Articles.showArticleHomePage&art_aid=94920

The key to the success of a viral campaign is emotional engagement. It is how the message or content is first perceived that causes us to want to pass it on. Often the more troublesome the message the faster it is shared; ElfYourself engages you with a feel good emotion and this is something you want to share with others. The Threshers voucher employed the fear that if you don't use it that week you will never see such an offer again. Engaging other emotions such as sadness, anger and disgust can also create an instinct to share content. Finally when planning your viral campaign ensure it is designed to help achieve your end goal.

Below is a basic diagram of the proliferation of a viral mail. The circles represent people. It is called viral as it spreads in a similar way as a virus.

8.12 Last word on tactics

Traditional marketing (TV, Radio, Magazine, Newspaper, Outdoor etc.) still plays a place in driving traffic online, The main difference is the cost, both in implementation and conversion. Traditional marketing is usually far more expensive than digital marketing and as such the conversion costs tend to be higher.

eg Typical Value Comparison

TRADITIONAL: Quarter page advert in HELLO magazine £6,435 - Circulation 427k.

DIGITAL: Full SMO campaign, reaching 15 social sites, 12 forums, two press releases, 100 social bookmarks, posts and profile management and reporting £5,500 per month - Reach 250m+

TRADITIONAL: Four Billboards for one month £3,200 low response and short lived

DIGITAL: £3,200 per month would fund SEO and Link building for long term high level exposure to specific audience

----

TRADITIONAL: Full page - firm day in The Sun £55,502, 8 million readers a day, possibly 0.1 - 0.2% traffic to site, £3.50 cpr

DIGITAL: £55,000, PPC at £1 per click 55,000 specific targeted visits

----

TRADITIONAL: 100,000 postcards printed and posted approx £30k 1-2% Response Rate - £15 cpr

DIGITAL: 100,000 Direct emails sub £10k 2-5% Response Rate - £5 cpr - £2 cpr

Ref: Typical Digital Marketing Agency

8.13 Web analytics

Nothing is more important than being able to measure and understand the success of your website. There are many aspects to analytics that can help grow and refine your website and even your business. Most analytics solutions involve adding a small piece of code to a web page.

There are many solutions for analytics of which the most notable are:

Google Analytics www.google.com/analytics
Webtrends www.webtrends.com

They all provide the same sort of information and a high level information segmentation. Key statistical information you get from analytics are:

Visitors: How many people came to your site and how extensively did they interact with your content? This traffic overview allows you to drill down and view the characteristics of different visitor segments and examine the different factors that make up visit quality (ie average page views, time on site and bounce rate).

Traffic Sources: Provides details on the different kinds of sources that send traffic to your site. "Direct Traffic" – visits from people who clicked a bookmark to come to your site or who typed your site URL directly into their browser. "Referring Sites" – visits from people who clicked to your site from another site. "Search Engines" – visits from people who clicked to your site from a search engine result page.

Content: Provides information on page view volume and usually lists the pages (Top Content) that were most responsible for driving page views. Entrance Points allows you to monitor the bounce rates for your most important landing pages. Consider redesigning any of the pages listed here that have a high bounce rate.

Goals: Goal conversions are the primary metric for measuring how well your site fulfills business objectives. A goal is a website page which a visitor reaches once he/she has made a purchase or completed another desired action, such as a registration or download. This information can also be displayed by an analytics tool.

Google Analytics. Available at: https://www.google.com/analytics [Accessed December 28 2008]

Google Analytics is a free tool and also considered to be the industry's best reporting tool.

It is also important to remember that many visitors to a website will use the telephone number on the site to make enquiries or purchases. If this is the case you should have some way of tracking where these customers came from whether it be simply by asking or by having a different telephone number displayed on a site depending on the referring URL. This can be done via programming and can help define how your marketing campaigns are doing.

> ### Assessment advice
>
> It would be worth going through the chapter again now that you have completed your initial reading to consider how each of the strategies and related tactics may be applied within your own organisations.
>
> Consider:
>
> What do we currently do?
>
> Would it be useful to implement anything else?
>
> What would be the actions points and implementation issues?
>
> How would we measure the outcome?
>
> It may take you a while to complete this activity. Don't worry, Chapter 7 is a very short chapter and simply summarises any remaining digital marketing theory. You will have additional time in Session 7 therefore to dedicate to this useful activity.

Discuss in detail a range of practitioner based digital marketing issues

- Growth in Internet marketing
- Defining requirements for a website
- Types of sites
- The importance of usability
- How to get a website built
- Digital strategies
- Digital tactics

BBC News Channel (Monday, 16 January 2006, 13:17 GMT) *'First Impressions Count for Web'.* Available at: http://news.bbc.co.uk/1/hi/technology/4616700.stm [Accessed 12 November 2008].

Chaffey, D., Smith, P.R., (2008). *Emarketing Excellence.* 3rd edition. Oxford: Butterworth-Heinemann.

comScore Releases May 2008 U.S. Search Engine Rankings. Available at: http://www.comscore.com/press/release.asp?press=2275 [Accessed 29 November 2008].

Email Marketing Reports (2008) *Why do email marketing?* http://www.email-marketing-reports.com/basics/why.htm [Accessed December 20 2008]

Hitwise UK - Leading Search Engines - October, 2008 http://www.hitwise.co.uk/datacenter/searchengineanalysis.php [Accessed 29 November 2008]

The Interactive Advertising Bureau (IAB). Available at: http://www.iab.net/iab_products_and_industry_services/1421/1443/1452 [Accessed December 16 2008]

Information Commissioner's Office, (2003) *The Privacy and Electronic Communications Regulations 2003.* Available at: http://www.ico.gov.uk/what_we_cover/privacy_and_electronic_communications/the_basics.aspx [Accessed December 20 2008]

Internet Advertising Bureau UK and PwC (2008) *£2.8 Billion Spent on Internet Advertising in 2007.* Available at: http://www.onlinecontentguide.co.uk/news-2009/2.8-billion-spent-on-internet-advertising-in-2007.html [Accessed 12 November 2008]

Microsoft (2008) Microsoft Junk E-Mail Filter Readme. Available at: http://office.microsoft.com/en-us/help/HA010450051033.aspx [Accessed December 20 2008]

Miniwatts (2008) www.internetworldstats.com. Miniwatts Marketing Group. http://www.internetworldstats.com/stats.htm [Accessed 12 November 2008]

TechCrunch. Available at: http://www.techcrunch.com/2008/04/29/end-of-speculation-the-real-twitter-usage-numbers/ [Accessed on December 14 2008]

YouTube.com (2008) Setting a Geographic Target for your Site in Webmaster Tools. Available at: http://uk.youtube.com/watch?v=r9r3PayqaZM [Accessed 20 October 2008]

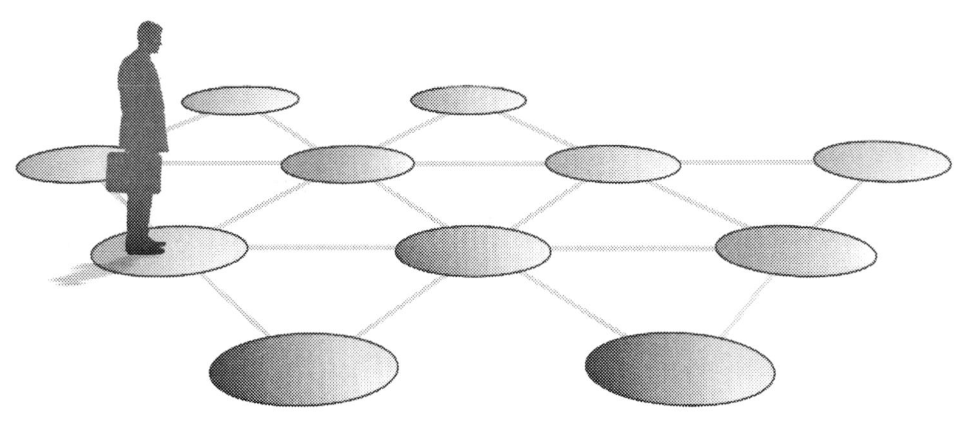

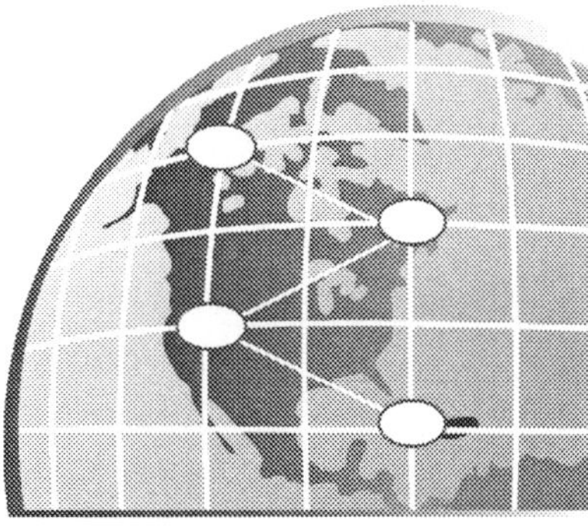

7 chapter

Integrating effective digital marketing strategies

In this particularly short chapter we bring together the academic and expert practitioner discussions about digital marketing as provided in the previous two chapters.

So far in Chapters 5 and 6, we have considered the nature of the digital market place and the implications the Internet has for the marketing mix. We have also looked at differences in online consumer behaviour and discussed in detail digital tools at the marketer's disposal. How to make your online presence more effective is covered from a practical perspective and so in this chapter we will draw together the threads to consider why and how these tools can be integrated with the wider communications strategy.

We also look specifically at the concept of the online value proposition – which actually should be different to the offline value proposition and highlight the benefits of an online presence.

Contents

Chapter learning outcomes

By the end of this chapter you will be able to:

- Evaluate the need for integrated digital campaigns

- Discuss the concept of the online value proposition

1 Integrating digital tools with the wider communications strategy

Damon Segal covered E-tools in Chapter 6. It is our intention in this chapter to discuss how these tools are integrated with the wider marketing communications strategy. Chaffey and Smith (2008) highlight the importance of mixing online and offline communications strategies. This is required so that:

- The **consistency of the brand message** is maintained

- Overall communications strategies are achieved through promoting a consistent message to achieve the same **objectives**

- To gain **maximum value and efficiency** from your online and offline communication activities

At the most basic level, all communications should carry where possible the URL for the related online website. Different customers prefer different communication channels and modes when making purchases, some will prefer to use online resources for research rather than purchasing for example. This means that when they refer to different brand messages to avoid confusion they should match.

Global case study

Coca-Cola, Lurpak butter, Walkers Crisps, Pepsi and many other brands successfully use their offline communication devises to help build traffic on their websites which in turn are used to build customer relationships via CRM and data collection devises.

Walkers for example ran a promotional campaign within the UK which offered free days out vouchers. This was communicated via:

- Traditional advertising via TV, radio and press

- The on-pack coupon and packaging design.

The coded coupons were on pack and could be exchange them for vouchers which could be exchanged with affiliate partners such as The National Trust, Holiday Inn and a number of theme parks and resturants. In order to 'bank' the coupons consumers were instructed to vsit the Walkers website which had a microsite specifically dedicated to the promotion.

The campaign used long standing Walkers celebrity endorser Gary Linekar and Cat Deely within ads and online as part of the central message.

A growing body of thought has recently emerged concerning media convergence.

1.1 Media convergence

> **Definition**
>
> **Media convergence** refers to bringing together digital devices so that information in different formats (audio, text, search, images, video) are accessed via one device (rather than several eg TV, radio, newspaper PC, mobile phone etc) (Fill, 2009).

An alternative view of media convergence is to consider different convergent dimensions:

- **Media technology** – digitisation has led to a reduction in the number of technological devises eg mobile phones such as the iPhone.

- **Media content** – almost all content can now be produced, edited, distributed and stored digitally

- **Media economics** – mergers of different media companies across sectors has lead to economies of scale

The ability to easily use direct marketing and customer relationship management more seamlessly though e-commerce and the use of these converged digital devices means that more than ever campaigns require consistency to avoid looking disjointed and as a result insincere. IMC provides the *opportunity* for organisations to integrate their communications and e-commerce has provided the *ability* to integrate communications (Fill, 2009).

Activity 1

Think about to what extent media convergence has impacted on your organisation from the perspective of being a seller and a purchaser of goods.

Fill (2009) Chapter 25 – Digital Media is a concise summary of digital technologies. A case study at the end of the chapter takes a detailed look at integrating traditional and digital media by a computer game brand.

1.2 An audience-centred approach

To be successful, Fill (2009) argues that IMC needs to adopt an audience-centred approach, one that accounts for the meaning that different audiences bestow on the messages they receive.

Global case study

MTV employed an integrated campiagn to encourage consumers to participate in a music video request show. To promote the show, artists were emailed, offline magazine ads in teenage girls' magazines ran, promotional postcards were distributed in schools and discussions on fan sites were held explaining the new programme concept.

The programme was aired live and asked consumers to vote via SMS text messages or via the Internet for the videos to be aired later in the show. Messages from voters were also shown while the winning videos were played.

In Chapter 8 we will look at the need for integrating marketing communications. In the meantime we shall refer to Pickton and Broderick's (2004) observations that the aim of marketing communications should be to:

- **Reach** the target audience
- Determine the appropriate **frequency** for messaging
- Achieve **impact** through creative for each media

By combining digital and traditional campaigns, there is an increased likelihood that the communications message will reach the target audience, more frequently (given multiple online and offline executions and therefore opportunities to see) and with more impact given the interactive capabilities of online media (Fill, 2009).

We will consider the concept of media neutral planning further in Chapter 9.

The Journal of Computer-Mediated Communication is available for free online at www.blackwell-synergy.com/loi/jcmc. The following article considers whether blogging has taken over the role of some traditional media channels: Meraz, S. (2009) *Is There* an Elite Hold? Traditional Media to Social Media Agenda Setting Influence in Blog Networks. Journal of Computer-Mediated Communication 14 (2009) 682–707. The article can be downloaded at: http://www3.interscience.wiley.com/cgi-bin/fulltext/122514480/PDFSTART

Activity 2

Think about how appropriate blogs are used within your own organsiation.

Identify three different audiences for the blog.

Why would these audiencies be interested in the blog?

The exact value of being online to the audiences is also an area that requires investigation.

2 Online value proposition

In the early days of the Internet, many so-called dotcoms failed. Generally they failed because their sites were too product-orientated and essentially were nothing more than complex product catalogues (Smith and Taylor, 2004). These failed websites didn't have any real, valuable proposition for their customers. Many were technology-driven, utilising whizzy web technologies but they didn't consider:

- how the technology helped consumers
- how to promote an attractive proposition
- whether there was a market need for the product or service proposition
- whether there was a niche or significant target segment which would sustain the business

Despite advocating the mixing of online and offline strategies, Chaffey and Smith (2008) argue that the **online value proposition (OVP) should be different to the offline proposition**. The unique advantages associated with being online include:

- immediacy
- interactivity
- depth of content
- faster and more convenient to buy online
- easier to make purchases
- cheaper prices
- more resources and information

The OVP should reinforce core brand values and summarise what a customer can get online that they cannot get elsewhere. Chaffey and Smith (2008) argue that most sites currently do not achieve this.

The authors suggest a number of OVP's which match their advertising straplines, here are a few of them:

- Autotrader – The biggest and best car site on the planet – www.autotrader.com
- Boosey and Hawkes – A world of music – www.boosey.com
- Easyjet – The web's favourite airline – www.easyjet.com
- Flickr – share your photos. Watch the world – www.flickr.com
- MUtv – The television channel dedicated to Manchester United – www.mutv.com

The OVP is more than the sum of features, benefits and prices and therefore should encompass the complete experience of selecting, buying and using the product or service.

2.1 The online customer experience

Delivering the online experience promised by a brand requires delivering:

- Rational values
- Emotional values
- Promised experience

There is therefore an inherent need to maintain service quality online (de Chernatony, 2001).

To provide exceptional customer experience, the following issues require consideration (Chaffey et al, 2009):

- ease of locating the site through search engines
- services provided by partners online on other websites
- quality of outbound communications such as e-newsletters
- quality of processing inbound e-mail communications from customers
- integration with offline communications

The following journal article written by GMN Faculty member Leslie de Chernatony will provide a good summary of branding in the context of the Internet:
de Chernatony, L., (2001). Succeeding with Brands on the Internet. *Journal of Brand Management*, 8 (3) pp. 185-95.

2.2 Permission marketing and the OVP

Building relationships is a delicate process. With the development of an online value which means something to consumers, they are more likely to give increasing levels of permission.

There are several steps to permission marketing:

- **Gaining permission** – permission to give customers information (usually via opt-in)

- **Collaboration** – in terms of both taking an active role – marketers help consumers buy and consumers help marketers sell by telling them what their needs are

- **Dialogue-trialogue** – dialogue emerges between 'relationship' partners via email, discussion rooms, focus groups, phone calls. Trialogue occurs between different consumers.

Relationships develop over time and there has to be respect, Godin (1999) summarised it with three words:

- Anticipated
- Relevant
- Personal

He continued to describe the essential components of permission marketing using a dating analogy, we have added our thought about the implications for IMC:

- **Offer an incentive to volunteer** (there has to be an excellent OVP to encourage this sometimes backed up with the promise of extra information or discounts)

- **Teach the consumer a little more about your product over time** (online the implications for this are that there should be fresh content or possibly new areas that consumers can opt into over an expended period)

- **Reinforce the incentive to guarantee that prospect maintains the permission** (minimise the likelihood that the consumer will want to opt out because they value the brand and relate to its values)

- **Offer additional incentives to get even more permission** (Lego for example has a Lego VIP club for customers who sign up to the website and shop within their stores)

- **Over time leverage the permission to change consumer behaviour towards profits** (loyal customers are more profitable, attempt to move customers through the relationship marketing loyalty ladder from prospects to advocates, database marketing enables marketers to identify who these customers are).

Chaffey and Smith (2008) provide some practical e-permission marketing guidelines as shown in the following table which they believe build on from Godin's original:

e-permission principle	application
Offer selective opt-in	Offer choice in communication preferences. Opt in is a legal requirement in many countries – four opt in options should be included: 1. content – news, products, offers, events 2. frequency – weekly, monthly, alerts 3. channel – email, direct mail 4. format – text or HTML
Create a common customer profile	A structured approach to data capture is required otherwise data will be missed and it will not be possible to compare between customers
Offer a range of opt-in incentives	Free-win-save incentives are common but different incentives are required for different audiences
Don't make opt out too easy	Although it is a legal requirement asking people why they want to unsubscribe provides useful CRM information. Setting up 'My profile' sections as on Amazon and Facebook helps
Watch – don't ask	Interruptive questions can be reduced by monitoring clicks to understand needs and trigger follow up conversations
Create an outbound contact strategy	Plan for the number, frequency and type of online and offline communications

Chaffey and Smith (2008) suggest that database marketing is at the heart of e-CRM and enables dialogue required for effective permission marketing which

- recognises and remembers each customer by **name and need**
- answers questions personally and sometimes automatically
- asks questions, collects information and builds a better profile
- delivers communications which are instantaneous, relevant and add value.

Assess your organisations approach to permission marketing.

Assessment advice

Within your assessment you may be asked to prepare something like a webinaire or webcast.

Webopedia.com defines Webcasting as "*[u]sing the Internet, and the World Wide Web in particular, to broadcast information*". *Basically a Webcast is a video program that is generally streamed over the Internet to your computer where you can watch it using appropriate software.*

Webcasts can consist of an video broadcast of a live presentation. The beauty of an archived broadcast is that you can stop the program, back up and review a section for clarity or greater understanding or because it's important and you want to fix an idea in your mind.

If you are unclear how these work in practice, it would be worth visiting the following websites for up to the minute guididance about the potential for these tools:

Mikogo is a free to use web conferencing provider – http://www.mikogo.com/guide/web-conferencing/

Advise about webcasts from the University of Washington –
http://depts.washington.edu/hswork/articles/Webcasts.html

Trying searching on Google using the term 'free Internet marketing webcast' – you are likely to come across a number of free example webcasts and it would be useful to famialiarise yourself with them to see how they work in action.

1 Evaluate the need for integrated digital campaigns

- Attention needs to be paid to the mixing online and offline communications strategies

- IMC provides the *opportunity* for organisations to integrate their communications and e-commerce has provided the *ability* to integrate communications

2 Discuss the concept of the online value proposition

- The online value proposition (OVP) should be different to the offline proposition

- The unique advantages associated with being online represent the OVP.

- Permission marketing, OVP and e-CRM are closely related.

1   There are likely to be several points. Even if you are not using digital technologies to sell products, the chances are your suppliers may well be and certainly your consumers will be familiar with some of the concepts and may have expectations for you to apply technology in the future.

2   This will depend on your own research.

3   This will depend on your own research.

Chaffey, D., Smith, P.R., (2008). *Emarketing Excellence.* 3rd edition. Oxford: Butterworth-Heinemann

Chaffey, D., et. al., (2009). *Internet Marketing: Strategy, Implementation and Practice.* 4th edition. Harlow: FT Prentice Hall.

de Chernatony, L., (2001). Succeeding with Brands on the Internet. *Journal of Brand Management*, 8 (3) pp. 185-95.

Fill, C., (2009). *Marketing Communications: Interactivity, Communities and Content.* 5th edition. Harlow: FT Prentice Hall.

Godin, S., (1999). *Permission Marketing.* New York: Simon and Schuster.

Pickton, D and Broderick, A., (2004). *Integrated Marketing Communications.* 2nd edition. Harlow: Prentice Hall.

Smith, P.R., & Taylor, J., (2004). *Marketing Communications: an Integrated Approach.* 4th edition. London: Kogan Page.

 Developing Integrated Communications Strategy

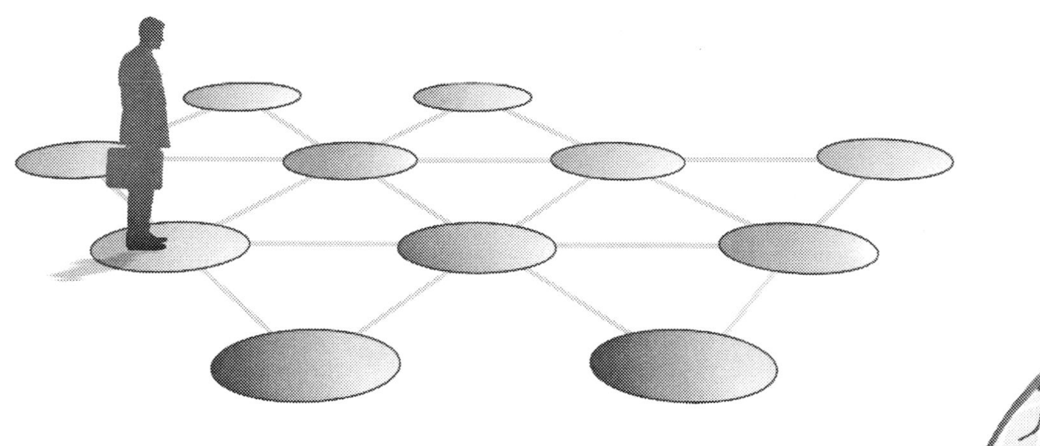
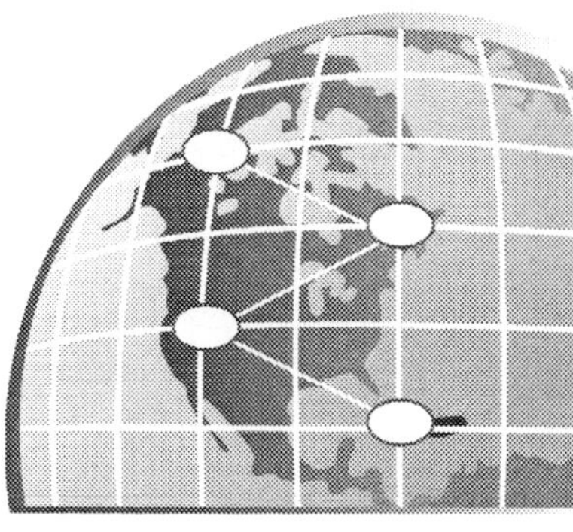

8 | chapter

The need for integrated marketing communications strategy

Within this chapter we look at the need to integrate marketing communications. Integration needs to occur:

- between the communications plan and the marketing plan

- between the marketing plan and the corporate plan

- within the communications plan in terms of delivering a consistent message in order to build relationships with stakeholders and promote a positive corporate reputation promoting consistency when communicating with wide ranges of internal and external audiences

The rewards of synergy and coherence are significant and this chapter takes each of the points above in turn to discuss the need for an integrated strategy.

The concept of relationships is heavily drawn upon in this chapter to show how relationship marketing, integrated marketing communications and corporate reputation are linked. The chapter begins by recapping the levels of strategy in operation within organisations.

Contents

In this chapter you will develop the skills to be able to:

- Discuss how communication strategy requires linking with the marketing and corporate strategies

- Introduce the concept of Integrated Marketing Communications

- Consider the role of stakeholder engagement and why relationships require management

- Identify the various internal and stakeholder groups and the importance of communicating with them

- Evaluate the need for strong corporate reputation

1 Communications strategy within the marketing strategy

1.1 What is integrated communications strategy (IMC)?

> **Definition**
>
> There are a number of alternative definitions of Integrated Marketing Communications (IMC). Here are two:
>
> **IMC** is *"a concept of marketing communications planning that recognises the added value of a comprehensive plan that evaluates the strategic roles of a variety of communications disciplines and combines them to provide clarity, consistency and maximum communications impact through the seamless integration of discrete messages."*(American Association of Advertising Agencies)
>
> *"IMC can represent both a strategic and tactical approach to the planned management of an organisation's communications. IMC requires that organisations coordinate their various strategies, resources and messages in order that they enable meaningful engagement with target audiences. The main purposes are to develop a clear positioning and encourage stakeholder relationships that are of mutual value."* Fill (2009)

The two definitions have several consistencies but also a key difference. It is quite typical to find **a range of alternative definitions** for marketing concepts and for IMC there appears to be more than for most (Pickton and Broderick, 2004). Pickton and Broderick (2004) summarise that most definitions contain the following elements:

- Clearly defined marketing communications objectives which are **consistent with other organisational objectives**

- **Planned approach** which covers the full extent of marketing communications activities in a synergistic way.

- Range of target audiences

- Management of **all forms of contac**t which may form the basis of communication activity.

- Effective management and **integration of all promotional activities** and people involved

- Incorporating **all product/brand and corporate communications efforts**

- Range of **promotional tools** – including personal and non-personal communications

- Range of messages – **brand positioning** should be derived from a single consistent message

- **Range of media** – any vehicle able to transmit marketing communication messages and not just mass media

Fill's (2009) definition further develops the concept in that he discusses the role in **encouraging mutually beneficial stakeholder relationships**. In Chapter 2 we looked at the concept of brand personality developing into brand relationships.

A common factor in each of the definitions highlights the need for IMC to be linked into the overall marketing strategy which is also linked to the strategic organisational plan.

The model below shows how the components of the marketing strategy should be combined in order to form the integrated communications strategy.

The integrated marketing communication process

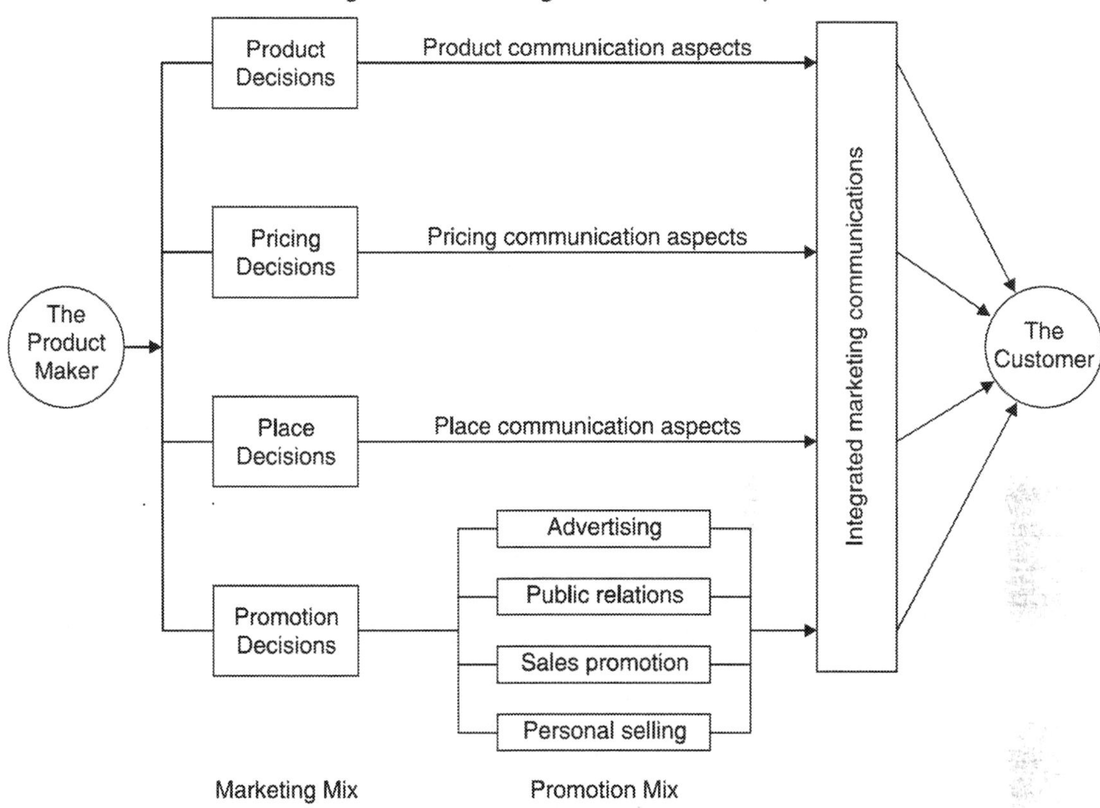

1.2 Benefits of integrated marketing communications

The principle benefit derived from integrating marketing communications is synergy (Pickton and Broderick, 2005). When for example According to Linton and Morley (1995) there are a number of benefits to developing an integrated plan, these are:

- Creative integrity
- Consistent messages
- Unbiased marketing recommendations
- Better use of media
- Greater marketing precision
- Operational efficiency
- Cost savings
- High-calibre consistent service
- Easier working relations
- Greater agency accountability

1.3 Disadvantages with taking an integrated approach

Fill (2009) identifies that IMC requires a change in thinking within organisations and changes in actions and expectations. As a result there can sometimes be resistance to integration. Related to this argument, although there are clear advantages to taking an integrated approach, disadvantages include:

- More centralised and bureaucratic procedures
- Potentially more management time seeking agreement from all parties
- Scope only for a uniform single message
- Creative opportunities may be limited due to standardisation
- Global brands may be limited in scope for local adaption
- A cultural change is required to overcome resistance
- If incorrectly managed there is the threat of severely damaging brand reputation
- Rarely are single agency networks available with access to all communication

1.4 Integrating communication with the corporate strategy

The nature of marketing communications strategy has to be viewed within the context of an overall marketing strategy, since a promotional strategy cannot exist in isolation. Prior to the promotional strategy being decided, the company will make a series of corporate and/or business unit decisions that will determine the nature of the overall marketing strategy. It is this which will then lay down the parameters within which a promotional strategy will be developed.

Each level of the organisation has a hierarchy of:

- Objectives
- Strategy
- Tactics

The tactics of the upper level then become the objectives of the next level down in the organisation. The levels we can usually consider are:

- Corporate (and then business unit, if separately managed)
- Functional (including marketing)
- Activity (including marketing communications)

The diagram below illustrates this '**planning hierarchy**'.

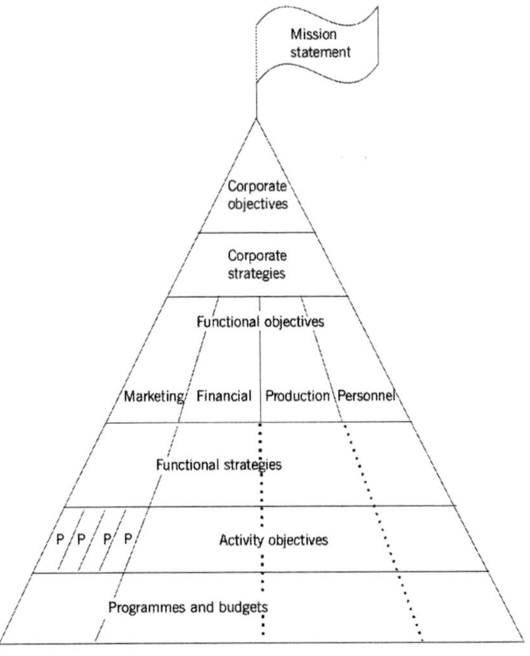

The planning hierarchy by Simon Majaro

An organisation's **mission statement** is a description of long-term vision and values. Mission statements have become increasingly common because they can provide clear guidance to managers and employees on the future direction of the organisation. In particular the mission statement can be used to develop the hierarchy of objectives that link the long-term vision and values with specific objectives at each level of the organisation.

In order to deliver an effective plan, it is important to establish **marketing communications objectives**. These will involve variables such as perception, attitudes, developing knowledge and interest or creating new levels of prompted and spontaneous awareness.

It should be clear from the diagram below that for our purposes there are three different forms of objectives: corporate, marketing and marketing communications objectives. Collectively these are referred to as promotional goals or objectives.

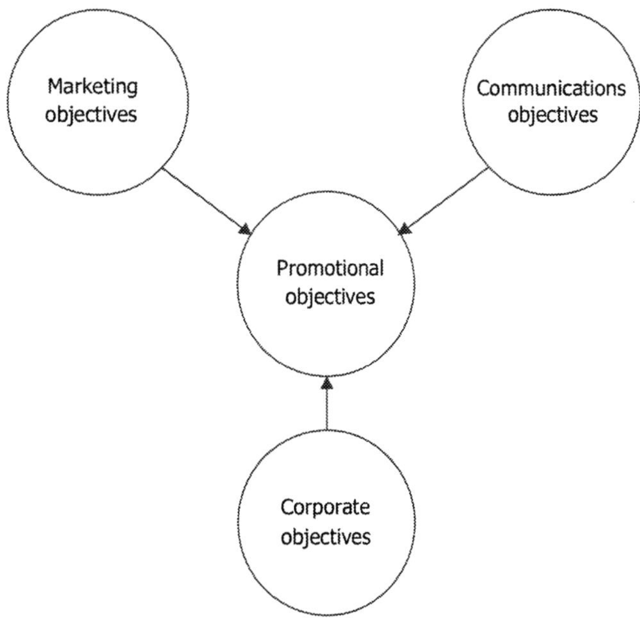

1.5 Communications objectives

Objectives need to be **specific** in that they must be capable of communicating to a target audience (who), a distinct message (what), over a specified time frame (when). Promotional objectives must therefore include:

- Identification of the **target audience**
- A **clear message**
- **Expected outcomes** in terms of trial purchase, awareness and so on
- A measurement of **results**
- Mechanisms for **monitoring and control**

The objectives need to be **measurable** and therefore quantifiable. Statements such as 'increase consumer awareness' are vague, whereas 'increase awareness of the 55 - 65 year age group from 40% to 80%' is more precise and capable of measurement.

Objectives need to be **achievable**. Purely from an internal company perspective, if sales are targeted to increase by 25% over a designated time period then manufacturing capacity will have to be secured to meet this target. Likewise, attempting to gain additional shelf space within a retail outlet will require that additional resources are devoted to the sales force, to sales promotions and to advertising.

Objectives need to be set with a degree of **realism** rather than on the basis of wild imagination. Otherwise, a company would be better off having no targets at all. An unrealistic target would tend to ignore the competitive and environmental forces affecting the company, the available resources at the company's disposal and the time frame in which the objectives have to be achieved.

Finally, objectives need to be **timed** over a relevant time period. Although a plan of action may be drawn up for a year, it will be the case that the plan will be reviewed against target, for example monthly or quarterly, so as to enable corrective action to be taken.

These principles of **SMART** (specific, measurable, achievable, realistic and timed) apply not only to the overall communication strategy but also to the setting of objectives for each tool within the promotional mix. Once the overall communication strategy has been set then individual (yet coordinated) plans need to be devised for each of the promotional tools. Using the SMART principle, objectives can be set for advertising, sales promotion, public relations, direct marketing, digital and Internet and personal selling. Each element of the communications mix should integrate with other tools of the communications mix so that a unified message is consistently reinforced (Smith and Taylor, 2004).

British Airways launched a sensational sales promotion when it announced free flights to anywhere in the world. The sales promotion was so newsworthy it hardly needed any advertising by the time the public relations people had maximised the editorial opportunities. The sales promotion had helped to build a database for future direct marketing activities (Smith and Taylor, 2004).

In recent years viral marketing has become very effective with similar strategies.

2 Stakeholder engagement

2.1 Stakeholder groups

There are many stakeholders within an organisation and various reasons for engaging them. Clearly not engaging with or building positive relationships with any stakeholder groups is a recipe for disaster. We will briefly review these groups to give you some ideas for analysing the nature of their stake or interest (what they want from the marketing organisation), and the nature of their influence and potential contribution (why the marketing organisation would want to take their interests into account, or develop relationships with them).

2.2 Categories of stakeholders

There are three broad categories of stakeholder in an organisation, as you may remember from your earlier studies.

- **Internal** stakeholders, who are members of the organisation. Key examples include the directors, managers and employees of a company – or the members of a club or association, or the volunteer workers in a charity. They may also include other functions of the organisation (eg marketing, production or finance) which have a stake in marketing activity, and/or separate units of the organisation (eg regional or product divisions) which have a stake in its plans.

- **Connected** stakeholders (or primary stakeholders), who have an economic or contractual relationship with the organisation. Key examples include the shareholders in a business; the customers of a business or beneficiaries of a charity; distributors and intermediaries; suppliers of goods and services; and financiers/funders of the organisation.

- **External** stakeholders (or secondary stakeholders), who are not directly connected to the organisation, but who have an interest in its activities, or are impacted by them in some way. Examples include the government, pressure and interest groups (including professional bodies and trade unions), the news media, the local community and wider society.

Here is a quick visual snapshot of the total stakeholder environment:

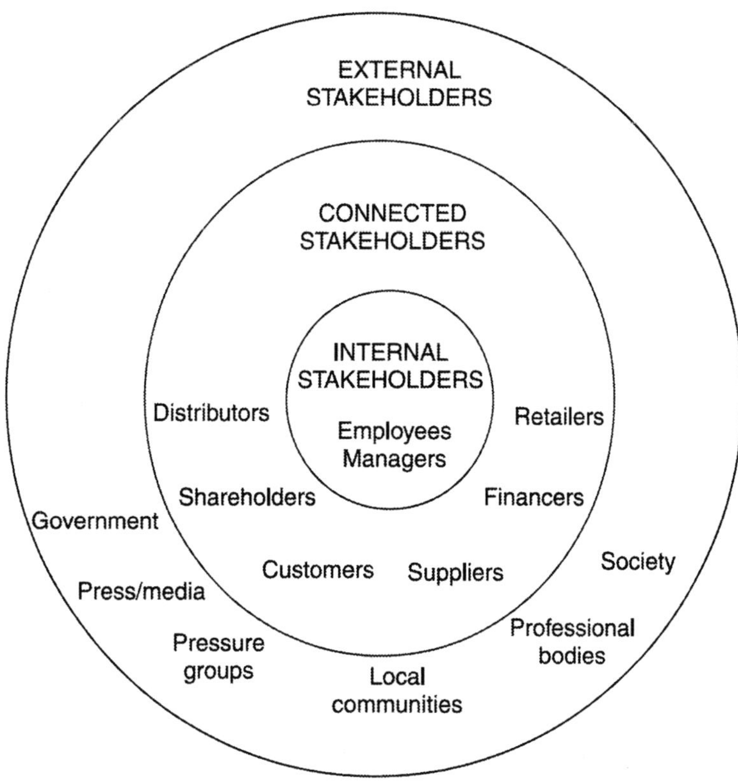

Stakeholders in the marketing organisation

While it is useful to categorise stakeholders in this way, it would be a mistake to think of them as entirely separate groups. Customers are also members of the wider community, and may be shareholders in the company and/or members of a consumer or environmental protection group, for example. Employees of the organisation may also be customers, shareholders and perhaps members of a trade union which can bring influence to bear on their behalf with the management of the organisation. So there are always areas in which membership and interests intertwine.

2.3 A stakeholder audit

Definition

A **stakeholder audit** is a systematic process of identifying stakeholders and assessing the effectiveness of current organisational strategies in relation to them.

A stakeholder audit is like a 'snapshot' of the state of the organisation's stakeholder relationships at a given moment. It may be accomplished, in the first instance, by informal brainstorming or discussion, followed up (if necessary) by more systematic research, using individual or focus group interviews or surveys.

The topics of a basic **stakeholder audit** might include:

- Identification of the main stakeholders in the organisation
- The needs/interests/concerns of each stakeholder
- The power/influence and potential impact on the organisation (positive or negative) of each stakeholder
- Current organisational strategies or thinking in relation to each stakeholder group
- Any problems or issues (perhaps reflected in particular incidents) arising from the handling of stakeholders, or a particular stakeholder

2.4 Stakeholder mapping

Mendelow (1985) developed a simple matrix to plot two factors for each stakeholder:

- How *interested* it is in influencing the organisation to get its needs met or interests protected (or in opposing or supporting a particular decision); and

- Whether it has the *power* to do so.

On the basis of these two factors, the matrix recommends the most appropriate type of relationship to establish with each 'quadrant' of stakeholder group:

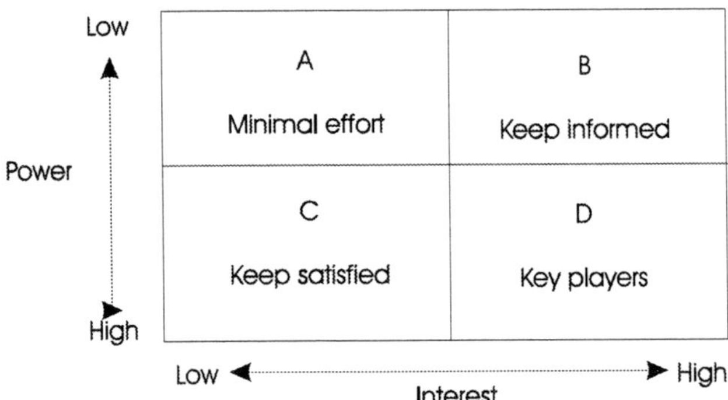

Mendelow's power/interest matrix

Working through each of the segments in turn:

- Stakeholders in **Quadrant A** who have *neither* interest in influencing the organisation/decision (because it doesn't impact on them greatly) *nor* the power to do so. They are a low-priority group: it will not be worth taking their goals into account, and they are likely simply to accept outcomes as they are.

  Small investors, or large suppliers with whom the organisation only does a small volume of business, may be in this category, as may local communities in relation to particular decisions with low immediate impacts. In relation to a marketing decision, other functions not directly affected by the decision may also be included here.

  The appropriate approach is to devote minimal effort to them – although they will need to be monitored in case their status changes.

- Stakeholders in **Quadrant B** are more important because of their high interest in influencing the organisation/decision. They may have low power to do anything about it, but unless they feel they are being kept 'in the loop' and understand the need for a strategy or decision, their concerns may lead them to seek additional power by lobbying or banding together against it.

  For the organisation as a whole, community, small supplier and employee groups may be in this category, in relation to decisions which impact significantly on their interests. For a marketing decision, small customers and staff affected by the changes may also come into this category.

  The appropriate approach is to keep them informed, and monitor and manage any issues that may arise.

- Stakeholders in **Quadrant C** are important because of their high power. They currently have little interest in using that power to influence the organisation, but if they become impacted, dissatisfied or concerned, their interest may be aroused and their power mobilised.

- A large institutional shareholder may be in this category, as may government agencies and regulatory bodies (in relation to organisations which are currently broadly compliant). For a marketing decision, senior managers in other departments, not currently significantly affected by the decision, may also fall into this category.

 Developing Integrated Communications Strategy

The appropriate approach is to keep such stakeholders satisfied, by ensuring that their needs are met and any concerns they may have are anticipated and addressed before they become 'issues' – without irritating them with excessive communication, which they may not see the need for.

- Stakeholders in **Quadrant D** are known as 'key players' they have high power and are highly motivated to use it in their own interests.

  Major customers, key suppliers and intermediaries, internal senior managers and strategic external allies/partners may be in this category.

  The appropriate strategy is to manage the relationship closely. This may include early involvement, consultation and negotiation, so that the key players' goals can be integrated with the marketing organisation's goals as far as possible. Plans must at least be *acceptable* to key players – and ideally, they can be encouraged to co-operate with the organisation, to mutual benefit.

In order to plan stakeholder management strategies in more detail, it may also be helpful to identify, on your matrix, which stakeholders are likely to be opposed to the organisation or particular plan (marked with a minus sign [-]) and which are likely to be supportive of it (marked with a plus sign [+]). This enables you to plan, for example, to engage the interest, or add to the power, of supporters – and to minimise the interest, or undermine the power, of opponents.

(a)   Complete a stakeholder audit for your own organisation.

(b)   Draw up a blank power/interest matrix, and place each stakeholder in the appropriate quadrant. For each, add a [+] or [-], according to whether they are supporters or opponents.

2.5 Stakeholder needs and relationships

The key point of stakeholder theory is that an organisation affects its environment and is affected by its environment. The boundaries of the organisation are highly permeable: influence flows from internal stakeholders outwards (eg through marketing) and from external stakeholders inwards (eg if a major customer pressures sales staff to represent its interests within the organisation, or more generally if a marketing-oriented organisation seeks to listen to its customers and meet their needs).

Note that stakeholders' influence is *not* just about power to get their needs met: stakeholders also make a positive contribution to the organisation's needs and objectives. (This is often what gives them influence: they have the power to give, or withhold, something the organisation wants.) Stakeholder management is ideally a mutual exchange of benefits – a classic marketing process and underpinned by strong foundations in relationship management theory!

> ### Definition
>
> **Relationship marketing** has been variously defined as:
>
> - 'The process whereby both parties – the buyer and provider – establish an effective, efficient, enjoyable, enthusiastic and ethical relationship: one that is personally, professionally and profitability rewarding to both parties' (Porter, 1993, p. 14)
>
> - 'A disciplinary framework for creating, developing and sustaining exchanges of value, between the parties involved, whereby exchange relationships evolve to provide continuous and stable links in the supply chain' (Ballantyne, 1994, p. 3)
>
> - 'The process of creating, maintaining and enhancing strong, value-laden relationships with customers and other stakeholders' (Kotler et al, 1999, p. 11)
>
> - 'Marketing based on interaction within networks of relationships' (Gummesson, 2002, p. 3)
>
> - 'All marketing activities directed towards establishing, developing and maintaining successful relational exchanges' (Morgan & Hunt, 1994, p. 22)
>
> - 'To identify and establish, maintain and enhance, and when necessary terminate relationships with customers (and other parties) so that the objectives regarding economic and other variables of all parties are met. This is achieved through a mutual exchange and fulfilment of promises.' (Grönroos, 2000, p. 242)

As you can see, definitions of Relationship Marketing vary in scope and emphasis. Some focus on buyer-supplier relationships and others on wider stakeholder networks. Some focus on the management of relationships, while others focus on their 'win win' nature – and this in turn is variously thought of in economic or wider social terms.

2.7 Principles and characteristics of relationship marketing

There are different schools of thought on the exact scope and nature of relationship marketing, and the term tends to be used to express a wide range of 'relationship-type' strategies. However, surveys of the literature (Christopher et al, 2002; Egan, 2004) highlight a number of distinctive features of Relationship Marketing as a broad philosophy and strategic approach.

- A shift from marketing activities which emphasise customer acquisition to those which emphasise **customer retention (in addition to acquisition),** with an intentional balance between the two – based on the economics of customer retention (discussed in section 4 below).

- The development of **on-going (and, if possible, constantly deepening and improving) 'lifetime value':** as opposed to one-off transactions. A key principle is to extend the duration, or lifetime, of a customer's purchasing relationship with the firm, and therefore to maximise their 'lifetime value': that is, the future flow of net profit arising from the relationship.

- Recognition of the potential for customer/supplier **co-operation, collaboration or partnership.** Traditional marketing could be seen as adversarial, with firms battling not only competitors, but customer bargaining power: if the customer won (say, on price or quality standards), the supplier lost – and vice versa. Relationship marketing is co-operative: by working together, customer and supplier can create, share and exchange more value, to mutual benefit (Gummeson, 2002; Grönroos, 1996).

This idea has been borrowed from industrial and service markets, where firms often have fewer, higher-value customers, demanding a more complex total value offering (product/service customisation, on-going service or consultancy, user training, collaboration on quality management, flexibility and so on) which requires long-term mutual investment and commitment.

- The aim of **long-term customer profitability**, or maximum lifetime value. Not all customers are equally profitable: the firm must use relationship *management* techniques to identify and prioritise potentially profitable customers – and avoid over-investing in unprofitable customers. Marketers may need to communicate in different (perhaps even contradictory) ways with customers and potential customers, depending on their status and worth to the firm.

- Emphasis on providing sustained and complex value to the customer, over and above the short-term satisfaction of product features and quality. This includes the recognition that:

    – High levels of service quality are required at every touch point with the customer

    – Quality, customer service and marketing are interdependent processes, and need to be more closely integrated

    – The marketing mix concept (4Ps) does not adequately describe all the elements which must be addressed to build and sustain relationships with key stakeholders: the service mix (particularly people and processes) is crucial

    – Customers are the ultimate definers of the value they wish to receive.

- A move from functionally-based marketing to **cross-functionally-based marketing**: customer value and quality are the responsibility of all employees, not just those who work in the marketing department. Internal marketing is recognised as critical in achieving external marketing success. (One co-founder of Hewlett-Packard is said to have remarked that 'marketing is too important to be left to the marketing department'!)

- The importance of **relationship values** such as trust, co-operation, commitment and mutuality (discussed in section 3 below). To have an on-going relationship, both parties need to trust each other and keep the promises they make: marketing moves from one-off potentially manipulative exchanges towards cooperative relationships built on mutual value exchange.

- The development of **supportive network relationships** with other internal and external stakeholders, rather than a focus on the customer-supplier dyad. Relationship marketing principles are extended to a range of diverse market domains (as in the Six Markets Model, see section 4), not just customer markets. Relationship marketing may even be defined by this full range of relationships, networks and interactions (as in Gummesson's 30R model – see section 4).

2.8 Marketing communications and customer relationships

Relationship marketing communications place a high emphasis on frequency, quality and personalisation of contact with customers and other stakeholders.

Multiple on-going customer contacts, using multiple *touch points* within the marketing organisation: eg sales or direct marketing, customer research and feedback-seeking, customer service, after-sales service/maintenance, the web site, loyalty programmes, newsletters, product up-dates, maintenance reminders, invitations to launches and other events, notification of special offers

Designated account managers or customer contacts may be used to focus initial contact on an individual touch point, to create personal familiarity and add value by having a single 'gatekeeper' to direct customer queries to other parts of the organisation.

Two-way dialogue with customers: not just marketing messages (business-to-customer or B2C) but customer to business (C2B) communication, through mechanisms such as feedback and suggestion seeking, the creation of customer communities and customer-generated web and advertising content.

This may be augmented by encouraging customer-to-customer (C2C) communication via discussion boards, user groups and customer networking events. C2C happens anyway, so it makes sense for the organisation to monitor the exchanges in order to gather information on customers' perceptions and interests), create a sense of belonging or affiliation (adding a social benefit to the total product/service offering) and offer social/entertainment value which draws people repeatedly to the web site and other mechanisms (where they can be targeted with promotional messages).

Personalised and customised contacts: making customers feel recognised and valued, and that their individual needs are being catered for. Examples include: customer 'recognition' by customer service staff (enabled by computer-telephony integration); the use of customer data to send birthday cards or service reminders (such as you might get from a dentist or car dealership); the personalisation of mailings, email and web pages; and the customisation of offers on the basis of customers' previous purchases.

3 Internal stakeholders

3.1 The importance of internal stakeholders

Organisations are made up of people (often referred to, these days, as the **human resources** of a business). When we talk about 'organisations' marketing to external customers/stakeholders, or establishing relationships with them, what we are really talking about is employees of the organisation implementing these activities.

Employees and their employing organisations are mutually dependent. Employees need work and its financial rewards in order to live – and organisations need employees to implement their plans and carry out their activities.

3.2 Internal marketing

Employees are therefore **key stakeholders** in the organisation, with both high interest and high (collective) power.

Peck et al (1999) identify the **internal market** as a key component of their Six Markets Model. It includes employees in all parts of an organisation with potential to contribute towards marketing effectiveness.

*"There are two key aspects to internal marketing. The first is concerned with how **staff work together across functional boundaries** so that their work is attuned to the company's mission, strategy and goals. The second involves the idea of the **internal customer**. That is, every person working within an organisation is both a supplier and a customer."* p. 302

Gummesson (2002) similarly suggests that:

*'The objective of internal marketing within Relationship Marketing is to **create relationships between management and employees, and between functions**. The personnel can be viewed as an **internal market**, and this market must be reached efficiently in order to prepare the personnel for external contacts: efficient internal marketing becomes an antecedent to efficient external marketing"* p. 198

Fill (2009) reflects on internal marketing from the perspective of internal communications when he advocates that:

"Internal marketing communications are necessary in order that internal members are motivated and involved with the brand such that they are able to present a consistent and uniform message to non-members." p. 897

We will explore these aspects in this chapter.

Definition

Internal marketing may be defined as a variety of approaches and techniques by which an organisation acquires, motivates, equips and retains customer-conscious employees (George & Grönroos, 1989), in order to help retain customers through achieving high quality service delivery and increased customer satisfaction.

Berry and Parasuraman (1991) define it as: 'attracting, developing, motivating and retaining qualified employees through job products that satisfy their needs' – which relates it clearly to the conventional concept of the marketing exchange.

Global case study

LL Bean (US catalogue retailer)

To inspire its employees to practise the marketing concept, LL Bean has for decades displayed posters around its office that proclaim the following:

"What is a customer? A customer is the most important person ever in this company, in person or by mail. A customer is not dependent on us, we are dependent on him. A customer is not an interruption of our work, he is the purpose of it. We are not doing a favour by serving him, he is doing us a favour by giving us the opportunity to do so. A customer is not someone to argue or match wits with, nobody ever won an argument with a customer. A customer is a person who brings us his wants; it is our job to handle them profitably to him and to ourselves."

3.3 The internal customer concept

As the term suggests, the internal customer concept implies the following ideas.

- Any unit of the organisation whose task contributes to the task of other units (whether as part of a process or in an advisory or service relationship) can be regarded as a supplier of a product/service. In other words, there is an **internal supply chain** – and the 'next person to handle your work' is your internal customer.

- The objective of each unit and individual thus becomes the 'efficient and effective **identification and satisfaction of the needs, wants and expectations of customers**' (one definition of marketing) within the internal value chain – as well as outside it.

- Any given unit of the organisation must 'create, build and maintain **mutually beneficial exchanges and relationships**' (another definition of marketing) within the organisation, as well as outside it.

3.4 Segmenting the internal market

The internal marketing mix (like the external marketing mix) will need to be adapted to the needs and drivers of the target audience. The internal market can (like the external market) be segmented to allow targeting to the distinctive needs of each group.

Jobber (2007) suggests segmentation of internal customers into:

- **Supporters**: those who are likely to gain from the change or plan, or are already committed to it
- **Neutrals**: those who are likely to experience both gains and losses from the change or plan
- **Opposers**: those who are likely to lose from the change or plan, or are traditional opponents

The product (plan) and price may have to be modified to gain acceptance from opponents. Place decisions will be used to reach each group most effectively (eg high-involvement approaches such as consultation meetings for supporters and neutrals). Promotional objectives will also differ according to the target group, because of their different positions on issues.

Christopher et al (2002) suggest an alternative way of segmenting internal customers, according to how close they are to external customers:

- **Contactors** have frequent or regular customer contact and are typically heavily involved with conventional marketing activities (eg sales or customer service roles). They need to be well versed in the firm's marketing strategies, and trained, prepared and motivated to service customers on a day-to-day basis in a responsive manner.

- **Modifiers** are not directly involved with conventional marketing activities, but still have frequent contact with customers (eg receptionists, switchboard, the credit department). These people need a clear view of the organisation's marketing strategy and the importance of being responsive to customers' needs.

- **Influencers** are involved with the traditional elements of marketing, but have little or no direct customer contact (eg in product development or market research). Companies must ensure that these people develop a sense of customer responsiveness, as they influence the total value offering to the customer.

- **Isolateds** are support functions that have neither direct customer contact nor marketing input – but whose activities nevertheless affect the organisation's performance (eg purchasing, HR and data processing). Such staff need to be sensitive to the needs of internal customers as well as their role in the chain that delivers value to customers.

Activity 2

Who are the internal customers of the marketing function in your organisation, or any organisation? How should they be managed?

We return to consider internal stakeholders and particularly the tools of internal marketing in Chapter 10.

4 External stakeholders

4.1 Identifying key external audiences

A number of influential models highlight the nature of a marketing organisation as a hub of relationships – and the positioning of marketing at the interface between the organisation and the other parties to those relationships.

4.2 The Six Markets model

The Six Markets model offers a helpful overview of the key categories of relationships for any given firm (sometimes called the 'core' or 'focal' firm, because we are looking at relationships from its point of view). It presents six role-related market domains or 'markets', each involving relationships with a number of parties – organisations or individuals – who can potentially contribute, directly or indirectly, to an organisation's marketplace effectiveness (Peck et al, 1999, p. 5).

The model has developed since its formulation in 1991, to take account of changing views and priorities in marketing, but the most commonly used version of the framework is shown below. Note that the focal firm is not the centre of the relationship 'hub', although the model recognises that internal marketing supports relationships with all the other parties. Rather, the customer is placed at the centre, 'to focus on

the purpose of relationship marketing: the creation of customer value, satisfaction and loyalty, leading to improved profitability in the long term' (Peck, et al 1999).

Customer markets

The concept of relationship marketing is based on the belief that firms must invest in building relationships with customers, in order to enhance profitability through customer retention and loyalty. The importance of customer relationships has long been recognised in professional and financial services, B2B marketing, and the market for regularly replaced consumer durables (such as cars). It is now 'catching on' in FMCG markets.

For consumer goods or services, the customer market domain represents **end customers**, users and consumers.

For B2B marketing, it also embraces channel **intermediaries**, including agents, retailers and distributors who are effectively 'customers' of the organisation, but operate between them and the end users.

Referral markets

Referrals, recommendations and endorsements by existing customers are an important source of new business: either directing potential new customers to the supplier (eg B2B sales 'leads' and professional referrals) or guiding consumer choice (eg through word-of-mouth recommendations or endorsements by trusted third parties).

Potential sources of referrals must be cultivated and motivated. 'Given that satisfied customers will happily endorse the products or services of the supplier if prompted, relationships with existing customers are an unrecognised or underutilised facility for many organisations' (Peck et al, 1999). Companies can create formal or informal cross-referral agreements between themselves and suppliers of complementary products (eg a weight loss consultancy and a local gym). Such referrals may also add value for customers, as part of a total service.

Internal markets

The internal market comprises all employees, and other functions, divisions and strategic business units (SBUs) of the firm. The concept of 'internal marketing', argues that employees and units throughout an organisation can contribute to the effectiveness of marketing to customers – most notably, through value-adding customer service and communications. It has been shown that employee satisfaction and retention (the aims of internal marketing) correlate directly with customer satisfaction and retention (the aims of customer relationship marketing) in service businesses (Schlesinger & Heskett, 1991).

Recruitment markets The recruitment market comprises:

- The **external labour pool**, and more specifically, those with the attributes and competencies needed by the firm, that is, quality potential employees.

- **Third parties**, such as colleges, universities, recruitment agencies and other employers, who can give the firm access to those quality potential employees. Relationships with these markets must

be cultivated in order for the firm to be able to compete with other employers to attract the best people, particularly in times, regions and disciplines in which there are acute skill shortages.

Influence markets

Customers' buying decisions are often made with input from a group of key influencers, referred to as a **'decision making unit' (DMU)**.

A range of **external third parties** also exercise influence over consumers – and over the marketing organisation itself. These influencers include governments and government agencies, the press/media, investors and pressure groups. Relationships with these markets can be exploited to generate positive PR (and/or minimise negative PR); influence public opinion in the organisation's favour; gain access to markets (eg through cause-related marketing); and enhance or replace other marketing activities (as in The Body Shop's exploitation of referral, media and pressure group relationships, in place of advertising).

'While relationships with these parties may not directly add value to a product or service, they can directly influence the likelihood of purchase or prevent an offer from even reaching the market' (Peck et al, 1999).

Supplier and alliance markets

The supplier market refers to the relationships that the firm must cultivate with its supply chain or network, in order to enable reliable, flexible, value-adding, cost effective flows of supplies into and through the firm to the end customer. The concept of supply chain management recognises the need for long-term, collaborative relationship development with a small number of suppliers, particularly for strategic or critical items – rather than hard-bargaining, adversarial, one-off transactions (which may still be used for routine items, where price is the main criterion).

The alliance market recognises a wide range of opportunities to add value through collaborative relationships between the core firm and partners (other than its immediate suppliers) in joint promotions, strategic alliances, joint ventures, knowledge sharing networks and 'virtual' collaborations.

Global case study

Coutts Bank: It's not just about clients

Coutts considers the way it services five distinct markets, in addition to its traditional client market, to make sure it maintains consistent, high quality relationships with them.

Internal markets: Coutts communicates with all staff – client account managers, product managers and support staff – about its relationship management priorities. The aim is to ensure there is no weak link in the chain that makes up the Coutts service offering.

Referral markets: Lawyers, consultants and financial advisers are a significant source of new business for the bank: they meet prospective clients every day and advise them on how best to invest their wealth. Coutts contacts these sources regularly and delivers regular, tailored information to them so that the bank is in their minds when they are advising their clients.

Supplier markets: Although the bank is a service provider, it needs to ensure that its tangible offerings – brochures, events, premises or staff lapel badges – match its service quality image. It works closely with a few suppliers who, over time, get to know its ways and standards.

Recruitment markets: In banking, new client account managers can often bring a portfolio of business with them, so Coutts works hard to sustain its quality image among its peers, and to be an organisation that people want to work for, in order to attract the best recruits.

Influence markets: One of Coutts' key influence markets is the governments and financial authorities in the countries in which it operates. They may actively seek the bank's views on legislative changes and new product opportunities that might attract investment to their countries.

4.3 Market relationships

Market relationships are the externally-oriented relationships between the suppliers, customers, competitors and intermediaries who operate in a market, which have traditionally been the focus of marketing. While seeking an alternative paradigm to replace the traditional marketing mix and focus on a relationship-orientated perspective, Gummesson (2002) identified 30 relationships which he believed operate but these have been condensed into

- Classic market relationships are the focus of traditional mainstream marketing management: the supplier/customer dyad; the triad or three-way relationship of supplier-customer-competitor; and the distribution network.

- Special market relationships focus in on certain aspects of the classic relationships, such as:

    - The interfaces between two parties: eg multiple contact points between suppliers and customers (especially in B2B marketing); and interfacing through full-time marketers (who directly create customer relationships) and other business functions (which indirectly influence them)

    - The various means through which parties interact: eg the service encounter (the interaction between a customer and front-line personnel), customer membership of loyalty programmes, and electronic relationships (interaction via IT networks)

    - The status and condition of relationships: eg distant and close relationships, and relationships with dissatisfied customers

    - The basis of relationships: eg relationship with objects (such as an i-Pod) and symbols (such as brands and corporate identity); law (contracts and compliance); non-commercial objectives (in the public/voluntary sectors and families); and green relationships (based on environmental and health issues)

4.4 Non-market relationships

Non-market relationships are relationships outside the market, but which indirectly influence the efficiency of the market relationships.

- Mega relationships are those which exist 'above' the immediate marketplace, in the economy or society in general.

These include:

- Personal and social networks, friendships and ethnic bonds, which often determine business networks (and can be exploited eg for word-of-mouth marketing and recruitment)

- Non-market networks (relationships with governments, legislators and influential individuals) and megamarketing (eg lobbying and public relations activity) to support marketing on an operational level

- Mass media relationships, which can be supportive or damaging to marketing

- Alliances (relationships and collaboration between companies) and knowledge relationships (since knowledge acquisition is a key reason for forming alliances)

- Mega-alliances (alliances beyond single companies, industries or nations: for example, the EU or North America Free Trade Agreement) which shape the macro environment of marketing

- Nano relationships exist 'below' the market relationships. They involve internal (intra-organisational) relationships, which may support or undermine the firm's external (inter-organisational) relationships and marketing. Examples of nano-relationships include:

    - Relationships between internal customers and internal suppliers in an organisation: how different tiers, functions and business units interact with one another as members of an internal supply chain

    - Concepts (eg total quality management) and organisational structures (eg matrix, product or account management) which can be used to build bridges between different functions, supporting integration and customer focus

    - Internal marketing: relationships with the 'employee market' which support external customer relationship marketing

5 Reputation management

5.1 What is corporate reputation management?

Reputation scholar Charles Fombrun offers an initial working definition of reputation as the sum of the images various constituencies (or stakeholders) have of an organisation:

Definition

'**Corporate reputation** is the **overall estimation** in which a company is held by its constituents. A corporate reputation represents the '**net' affective or emotional reaction** – good-bad, weak or strong – of customers, investors, employees and general public to the company's name' (Fombrun, 1996, p. 9).

The reputation of an organisation – and indeed that of a person or a place – is a kind of aggregate of the perceptions built up in the minds of interested parties: how others 'think of you' or your character or standing, as an overall assessment. Varey (2002) suggests that:

"Corporate reputation is an all-encompassing term for what employees think about their employer, what customers think about their provider, what investors think about their shareholding and so on." p.193.

As we will see, reputations are built up over time, as people have positive and negative direct experiences of the organisation, and receive positive and negative messages from it or about it – each reinforcing or adjusting their overall evaluation. Reputations are therefore built on both the reality of what an organisation is (or how people experience it), and the messages that are conveyed by and about the organisation, both of which shape the perceptions of its various stakeholders. Doorley & Garcia (2007, p. 4) express this by the simple formula:

Reputation = Sum of Images = (Performance and Behaviour) + Communication

As Varey (2002) argues:

"Reputation does not originate from the corporate communications office, or the marketing plan, or individual behaviour. It is not a fabricatable artefact that can be used to manipulate others' feelings, in the way that advertising can be (mis)used. Reputation springs from experiences, thought processes and values of people who see themselves as stakeholders of a business." p. 193

We should emphasise that reputation management requires attention to both the reality of performance and behaviour and the messages conveyed by corporate stakeholder communication and stakeholders' communications among themselves. It is not just about marketing-controlled 'spin'!

Quick 'spot check' to focus your thinking on corporate reputation. In two or three words, or a short sentence, write down your answers to the following questions.

* What kind of employer is your work organisation?

* What do you think of your banking services provider?

* If you own shares in a company, what is your overall impression of your investment?

* Where would you want to go on your next overseas holiday, and why?

* If you had a significant sum of money to donate to charity, which charity would you choose and why?

* Is there an organisation that you would definitely not do business with? If so, why?

5.2 The benefit of a positive reputation

Broadly speaking, a positive reputation is a key source of distinctiveness for an organisation, which can differentiate it from its competitors; produce support for and trust in the organisation and its products (which in turn may enhance its stability). Fill (2009) outlined four reasons why a strong reputation is considered to be strategically important:

1   Differentiation – especially where there is little difference at product level
2   Effect on the share price
3   Support during turbulent times as a measure of corporate value
4   Higher quality customer relationships

Watson & Kitchen (2008) cite the following case vignette of **Johnson & Johnson**, to illustrate how reputation speeds recovery after a crisis.

'The case study of the Johnson & Johnson Company and the pain relief drug Tylenol is one most frequently referred to. Although it happened two decades ago, it illustrates the value of a strong brand image, a strong company reputation, and an ethical core to the business which was immediately operationalised once a problem arose. Note that the managerial response to this case is a response which few firms seem to carry out nowadays. A few capsules of Tylenol were contaminated with cyanide by an unknown person and the top-selling analgesic was immediately withdrawn from the market. Market research showed a high level of confidence in both the Johnson & Johnson name and the Tylenol brand, and this trust was used as the basis to relaunch the drug after the packing had been redesigned to make it tamper-proof. The company's speed of response and highly effective communications throughout the crisis process – closely allied to its decision to take its best-selling product off the market – has become the management template for product withdrawal and ethically prompt corporate communication.

Jim Burke, then chairman of Johnson & Johnson, said *'the reputation of the corporation, which has been carefully built over 90 years, provided a reservoir of goodwill among the public, the people in the regulatory agencies, and the media, which was of incalculable value in helping to restore the brand'* (quoted in Dowling 1994, p. 215). Even now, Johnson & Johnson continually heads the list of most respected corporations in the USA.'

For reflection: What features of Johnson & Johnson's corporate character, practice and communication contributed to its resilience in the face of crisis? (You may like to pick out some key phrases from the account given.)

Claims for the link between reputation and business performance are controversial. They are often supported by rankings such as America's Most Admired Companies surveys – but the reputation measure used in such surveys includes an appraisal of the financial performance of the company, so a correlation is hardly surprising. In fact, it may be less a case of reputation leading to financial success, than a case of financial success creating a positive reputation, since this factor is most likely to 'impress' the survey respondents (peer-group managers) and influence their perception of other measures (the 'halo effect').

If you would like to explore a range of reputation factors, and their link to business performance, via the America's Most Admired Companies study, see: http://money.cnn.com/magazines/fortune/mostadmired/2009.

It is worth noting that in certain market conditions, companies that have what might be regarded as a negative image (being aggressive, being a poor employer, exploiting workforce or suppliers, damaging the environment) – still survive and prosper. Examples might include Wal-Mart in the United States, or the continuing marketing success of companies like McDonalds (in view of reputational damage due to alleged damage to health and the environment), Nike (alleged exploitation of workers in low-cost labour countries), Shell (environmental damage), Perrier (highly publicised contamination fiascos) and Microsoft (trouble with competition regulators).

5.3 The Reputation Value Cycle

van Riel & Fombrun (2007) describe how financial value and stakeholder support are dynamically related, in their **Reputation Value Cycle**: see below.

"Endorsements build value, and enable a company to expense [sic] funds on corporate activities such as advertising, philanthropy and citizenship that generate media endorsements, attract investors, and add financial value. The need effect is a reinforcing loop through which communication, recognition, endorsement and supportive behaviours from stakeholders create equity and financial value." p. 271

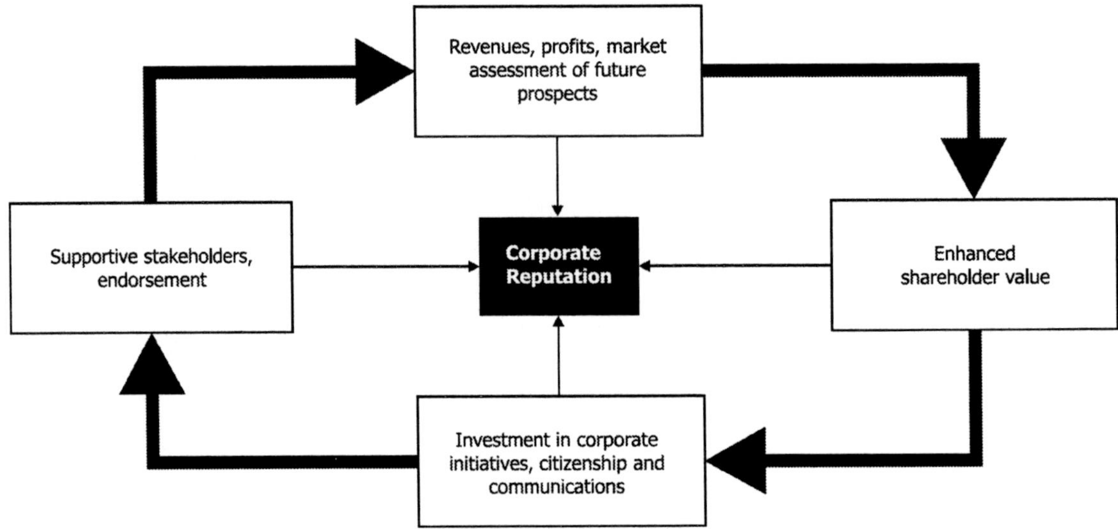

5.4 Creating a strong, sustainable and positive reputation

As we will see, strong reputations are partly formed by corporate communications designed to produce favourable perceptions in the minds of the public (by corporate '**expressiveness**'). Research by Fombrun and van Riel (2004) suggests that the expressiveness profile most supportive of a strong, sustainable and positive reputation is made up of five elements.

- **Visibility**. High reputation companies invest in communications.

- **Distinctiveness**. High reputation companies differentiate themselves (via the identity mix) from their competitors.

- **Authenticity**. High reputation companies match their communications to their values and actions, and live up to their promises and commitments.

- **Transparency**. High reputation companies have a culture of openness, having 'nothing to hide'.

- **Consistency**. High reputation companies cultivate consistency and coherency in the messages given by their different business units and brands, and avoid 'mixed messages'.

Global case study

Fill (2009) suggests that Fombrun's criteria for a favourable reputation can be illustrated with reference to a company such as **Nokia**, the Finnish mobile phone manufacturer.

'**Credibility** is established through its range of products, which are perceived to be of high quality and branded.

Trustworthiness has been developed through attention to customer service and support. **Reliability** and **consistency** have been achieved by setting and adhering to particular standards of quality, and **responsibility** is verified through a strong orientation to service and values manifested through the company's strong product development and innovation policy.'

5.5 A The stakeholder perspective on reputation

Different stakeholder groups may have different expectations, satisfiers and perspectives in relation to an organisation. Broadly:

Employees may be looking for an employer they can trust, to provide an on-going livelihood, to uphold their employment rights, and not to exploit their labour and contribution.

Customers may be looking for a reliable provider of quality products and services, reducing their risk in purchase decisions by the consistency with which the company fulfils their needs and expectations.

Investors may be looking for a credible company, reducing their investment risk by the confidence inspired by the company's management, track record and market standing.

The community as a whole may be looking for a responsible business, which takes into account its potential impacts on people and the environment, and invests in good corporate citizenship.

(Davies et al, 2003)

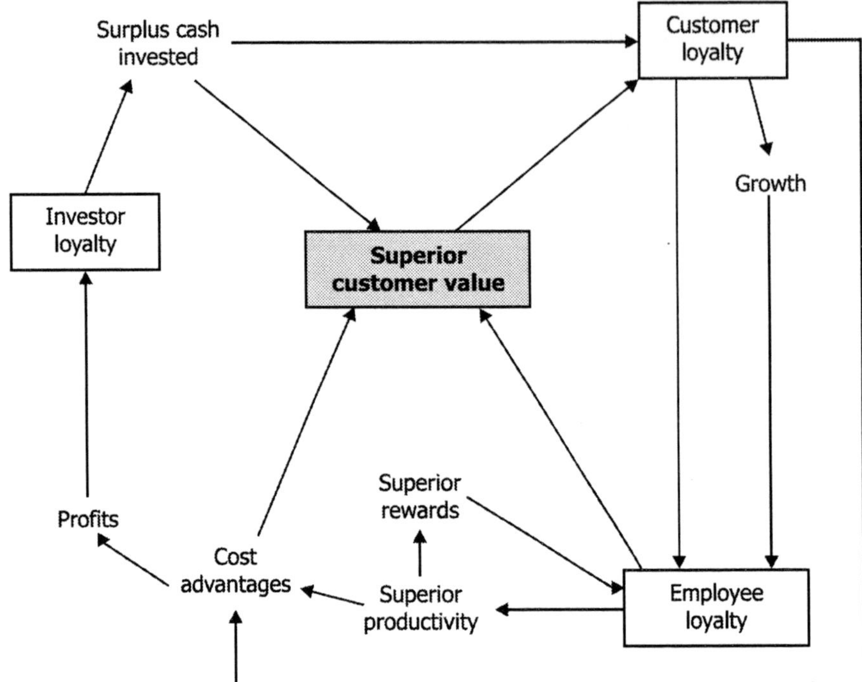

5.5.1 Credibility

> **Definition**
>
> **Credibility** is the extent to which corporate statements, promises and claims are believable to investors (and other stakeholders): the belief that the organisation will do what it says it will.

An organisation, and by extension its communications, may be credible by virtue of factors such as:

- **Track record and credentials**: proven and consistent performance, attainments, capabilities and capacities; proven and consistent record of following through on claims and promises.

- **Market and industry standing**: reputation among peers; market/industry awards; the quality of suppliers and other business partners associated with the company.

- **Character**: demonstrated (or a reputation for) integrity, honesty and good intent.

- **Credible leadership**: visible, expressive, high-status, peer-respected leaders; competent management; strong (innovative, sustainable) strategic management and direction.

- **Credible spokespersons or intermediaries**: eg recognised and respected authorities, co-opted to present or corroborate the organisation's message. (You may recognise this from 'scientific' and celebrity advertising and endorsements, for example.)

- **Objectivity**: demonstrated consideration of both sides of an issue (or stakeholder conflict) before formulating the organisation's approach or perspective.

- **Coherence and consistency** of messages (from different leaders, business units and brands) and between messages and behaviours/results.

- **Asset backing** for claims and promises requiring investment.

- **Sound risk management** practices, supporting business continuity and stability.

> **Definition**
>
> **Trustworthiness** is closely allied to credibility, describing how far stakeholders are exposing themselves to risk in dealing or allying themselves with the company: in other words, the likelihood that their expectations will be disappointed or that they will be 'let down' by the company.

An organisation may make itself trustworthy (worthy of trust) by virtue of factors such as:

- Track record of consistently delivering on promises or claims made, or expectations raised. (Note that the management of expectations – eg avoiding 'over-claiming' – is just as important as fulfilling expectations in building trust: 'under promise, over deliver' is a good motto...)

- Demonstrated (or strong reputation for) honesty, integrity and fairness, which would support the belief that the company will 'do the right thing' by its stakeholders.

- Consistent reliability of performance, service and brand qualities: building stakeholder confidence that they know what to expect.

- Credibility of expressed values, such as customer focus, value for employees, quality or environmental sustainability – which increases the stakeholder's belief that the company will 'do the right thing' if issues in these areas arise.

- Transparency: willingness to share information appropriately, in order to support stakeholder decision-making and collaboration. Transparency also demonstrates trust, because information can be misused – used to the advantage of one party at the other's expense (eg exploiting information on a supplier's costs or problems to strengthen one's bargaining position in price negotiation) or released to unauthorised third parties (eg giving customer data to commercial mailing lists, leaking unfavourable company reports to the press, or divulging plans to competitors).

- Relationship marketing: strategies aimed at on-going contact and connection with customers and other stakeholders, rather than one-off transactions. Trust is built up by experience – the more often and consistently stakeholders have positive experiences and encounters with the company, the greater the level of trust will be.

The requirements for developing trust have been described as Three Cs:

- Competency. Does the organisation know what it is talking about? Can it deliver on claims and promises?

- Caring. Does the organisation asking for trust really care about the issue and its stakeholders? (If so, it is more likely to be fair and supportive in dealing with them.)

- Character. Does the organisation have character and values (eg honesty and integrity), on which it consistently bases its behaviour?

'If there is a vacancy in any of these areas, all is not lost: consider filling in the gaps. It may be a matter of who is chosen as the spokesperson to deliver a message, or it may involve drafting a recognised expert to corroborate the message or appeal. People will respect competence, disregard incompetence, and respect the incompetent when they are honest about their lack of competence. Begin with a firm grip on reality, build small wins, and advance trust one step at a time.' (Davies et al, p. 280).

5.5.3 Reliability

Again, reliability is closely related to credibility and trust. Can stakeholders believe in the organisation (credibility)? Is the organisation worthy of that trust (trustworthiness)? Can stakeholders therefore rely on the organisation and take a risk on purchase or relationship, without having their rights or legitimate expectations disappointed?

For customers, reliability is crucially a matter of consistency of product and service quality. A reputation for reliability depends on the organisation's ability consistently to fulfil customer expectations and to create a positive experience of doing business with it, at every encounter and touch point with the organisation.

Samuel (2007, p. 63) emphasises that 'No matter how much effort we spend on promoting, positioning or profiling ourselves, failing to live up to individual expectations will severely damage our reputation. Everything counts. Every element of every transaction either enhances or damages our reputation. Whatever statements or claims we make, we must deliver on. No excuses. No blame. At all costs we must avoid giving others the opportunity to doubt us, as doubt calls our reputation into question.'

- Modern consumers want instant gratification – and alternative providers and solutions are available if one organisation fails to get it right first time.

- Reputation depends on the 'weakest link': the potential for the organisation to fail to deliver customer expectations at any interaction or touch point.

- Doubt spreads from a single demonstration of unreliability to damage reputation: 'If they don't deliver in this area, what else might they fail to deliver on?'

- People seek reinforcement and confirmation of their expectations. If they believe they have been failed in some way, they are more likely to seek out and focus on information that would tend to confirm this belief: negative experiences of others, negative press coverage and so on.

Service encounters are potentially critical incidents in building, maintaining or losing a reputation for reliability. A single disappointing service encounter – and/or a firm's subsequent poor response to handling the problem may be sufficient to reduce loyalty, make the customer more amenable to switching brands in response to competitor offers, and create reputation-damaging negative word-of-mouth. Customers may tolerate negative critical incidents for a while, but they are taken into account in the long-term evaluation of the organisation's performance.

Nor is this dynamic confined to customers. Think, for example, of the effect on a supplier's willingness to deal with an organisation, if it is unreliable in the payment of its debts or following through on agreed actions.

5.5.4 Responsibility

> **Definition**
>
> The concept of **corporate social responsibility (CSR)** embodies a range of responsibilities that an organisation might consider it has towards its wider or secondary stakeholders, including the society and natural environment in which it operates.

Carol & Buchholtz (2000) suggest that there are four main 'layers' of corporate social responsibility.

- **Economic responsibilities**. The firm must produce goods and services wanted by the market, profitably – otherwise it cannot survive, and will not be able to fulfil any other obligations to stakeholders. Economic responsibilities include: operating efficiently and effectively, aiming for consistent levels of profitability, and competing effectively in the market.

- **Legal responsibilities**. Economic goals must be pursued within the framework of the laws of the society within which the firm operates. It is sometimes said that 'the law is a floor': it does not define best practice, but does set out the minimum principles and standards that are considered acceptable. The news is full of examples of how business practices can fail to comply with contract, employment, consumer protection, health and safety, competition law and so on – and the consequences of non-compliance.

- **Ethical responsibilities**. Ethical responsibilities comprise the expectations of society, over and above basic economic and legal requirements: honesty in business relationships, for example; not exploiting small suppliers and retailers; or not marketing manipulatively to children. This is a more discretionary area, because there are fewer direct sanctions against unethical behaviour. Ethical principles may be enforced via Codes of Ethics or Practice in an industry, profession or individual firm – but they are often also subject to stakeholder pressure (demands, protests, boycotts and so on).

- **Philanthropic responsibilities**. Above and beyond even ethical dealings, society increasingly expects that marketing organisations be 'good corporate citizens': that is, that they proactively and positively contribute to the society in which they operate. Examples of philanthropy include corporations building community amenities, sponsoring local causes and events, donating money to charity, promoting or campaigning on issues of concern and so on.

Society as a whole, because of its diversity, may have low direct influence on the policies and activities of an organisation.

However, society's interests are organised, focused and represented in various ways: by government policy, legislation and regulation; by 'consumerism' or the consumer rights movement; by pressure and interest groups seeking to exert influence on behalf of particular constituencies or on particular issues; and by the fact that wider society is part of the environment within which the organisation operations – and within which it competes for labour, suppliers, customers, support and other key resources.

Global case study

Alcoholic versions of soft-drinks – so-called 'alcopops' – have caused controversy and opposition worldwide, particularly because the brands seem to be directed mainly at young people, in the face of widespread concern at the level of youth alcohol abuse and binge drinking.

The need for tighter controls over the marketing of alcopops has been recognised by the drinks industry itself. A code of practice was introduced by the **Portman Group**, an organisation founded by the major UK drinks producers to promote sensible drinking and to reduce misuse of alcohol.

The code of practice complements and is consistent with all other relevant self-regulatory codes, and it helps to control the industry without the burden of new legislation. The provisions of the code are wide-ranging and cover the naming, packaging and merchandising of drinks.

An example of its operation could be seen in the marketing of Carlsberg-Tetley's 'Thickhead' drink, which was held by the Portman Group to be breach of industry guidelines, and a 'serious misjudgement'. Carlsberg-Tetley responded by agreeing to change the label.

Activity 5

What other industries and entities (eg sporting codes) have taken concerted self-regulatory action to change their culture or practices, in order to improve or protect their reputations?

Watson and Kitchen (2008) cite Fombrun when they quoted:

'Better-regarded companies build their reputations by developing practices which integrate social and economic considerations into their competitive strategies. They not only do things right – they do the right things. In doing so, they act like good citizens. They initiate policies that reflect their core values; that consider the joint welfare of investors, customers and employers, that invoke concern for the development of local communities; and that ensure the quality and environmental soundness of their technologies, products and services.' (Watson & Kitchen, op cit, p. 126).

(In turn, trust is central to the success of relationship marketing strategies, as it reduces the perception of risk and supports mutual investment in relationship. Without trust, the investment of time, money and commitment in a relationship, on either side, would simply be too risky: there would be no reason to believe that the benefits, promised in exchange, would accrue.)

1 Discuss how communication strategy requires linking with the marketing and corporate strategies

- A hierarchy of planning should exist with corporate, functional, and activity plans.

- Mission statements can be used to develop the hierarchy of objectives

- Corporate, marketing and communication objectives form promotional objectives

2 Introduce the concept of Integrated Marketing Communications

- Integrated marketing communications has a number of definitions but generally it involves an integration of the entire marketing mix in order to encourage mutually beneficial stakeholder relationships through co-ordinated communications to deliver a clear and consistent message.

3 Consider the role of stakeholder engagement and why relationships require management

- Stakeholders give loyalty in exchange for their expectation that value will flow to them from a relationship with the organisation.

- Relationship marketing advocates the need for long term relationships in order to achieve long term strategic prosperity.

- Relationship marketing communications place a high emphasis on the frequency, quality and personalisation of contact with customers and other stakeholders.

4 Identify the various internal and stakeholder groups and the importance of communicating with them

- Groups include internal, external and connected stakeholders.

- Audits and stakeholder mapping provides a snapshot if the relationships in a given market and may assist with communications planning.

5 Evaluate the need for strong corporate reputation

- Positive reputation is a source of distinctiveness for an organisation which can differentiate it from its competitors; product support for and trust in its products.

1   This will depend on your own research. Try to think broadly and add as many stakeholders as you can. You should be able to apply this within your assignment as it is useful preliminary research.

2   This will depend on the nature of your organisation.

3   This will depend on your research.

4   Features include the company's swift action working with the media and regulatory agencies and established goodwill.

5   There are many and it will depend on individual countries but they include advertising, food manufacturers, health practitioners, motor industry, energy industries, packaging manufacturers and many others.

AAAA (2009). *Definition of Integrated Marketing Communications.* Available at www.aaaa.org.

Carroll, A. B. & Buchholtz, A. K., (2000). *Business and Society: Ethics and Stakeholder Management.* Cincinnati: South-Western College.

Christopher M.G., Payne A.F. & Ballantyne D., (2002). *Relationship Marketing: Creating Stakeholder Value,* Oxford: Butterworth-Heinemann.

Davies G., Chun R., Da Silva R .V. & Roper S., (2003). *Corporate Reputation and Competitiveness.* Abingdon, Oxon: Routledge.

Dowling, G. R., (1994). *Corporate Reputations: Strategies for Developing the Corporate Brand.* London: Kogan Page.

Egan, G., (1994). *Working the Shadow Side: A Guide to Positive Behind-the-Scenes Management.* SF: Jossey-Bass.

Fill, C., (2009). *Marketing Communications: Interactivity, Communities and Content.* 5th edition. Harlow: FT Prentice Hall.

Fombrun, C J., (1996). *Reputation: Realizing Value from the Corporate Brand.* Harvard Business School Press: Boston.

Fombrun, C J & Riel, C .B. M. van, (2004): *Fame and Fortune: How Successful Companies Build Winning Reputations.* Upper Saddle River, Pearson Education: NJ.

Gummesson, E., (2002). *Total Relationship Marketing.* 2nd edition. Oxford: Elsevier Butterworth-Heinemann.

Grönroos, C., (1996). Relationship Marketing: Strategic and Tactical Implications in *Management Decisions,* Vol 34, no 3, pp. 5-14.

Jenkinson, A. and Sain, B., (2005). *Harley Davison: Organisation-Led Integrated Marketing.* The Sparkles Series, The Centre for Integrated Marketing, University of Luton.

Jobber (2007). *Principles and Practice of Marketing,* 5th edition. Maidenhead: McGraw-Hill Education.

Kotler P., (1997). Method for the Millennium in *Marketing Business.* February pp. 26-27.

Kotler P., Armstrong G., Meggs D., Bradbury E., and Grech J., (1999). *Marketing: An Introduction.* Sydney: Prentice Hall Australia.

Mendelow, A., (1985). Stakeholder Analysis for Strategic Planning & Implementation in *Strategic Planning & Management Handbook,* King & Cleland (eds). NY: Van Nostrand Reinhold.

Morgan R.M., & Hunt S.D., (1994) The Commitment-Trust Theory of Relationship Marketing in *Journal of Marketing,* Vol 58, no 3, pp. 20-38.

Peck H.L., Payne A., Christopher M. & Clark M., (1999). *Relationship Marketing: Strategy and Implementation,* Oxford: Elsevier Butterworth-Heinemann.

Pickton, D and Broderick, A., (2004). *Integrated Marketing Communications.* 2nd edition. Harlow: Prentice Hall.

Porter, C., (1993) quoted in The Marketing Strategy Letter, May 1993, p. 14.

Samuel, H., (2007). *Reputation Branding.* Wellington, NZ: First Edition Ltd

Smith, P.R., & Taylor, J., 2004. *Marketing Communications: An Integrated Approach.* 4th edition. London: Kogan Page.

Varey, R. J., (2002). *Marketing Communication: Principles and Practice.* Abindgon: Routledge

Watson & Kitchen Watson, T. & Kitchen, P. J., (2008). Corporate Reputation in Action in Melewar T C (ed.) *Facets of Corporate Identity, Communication and Reputation.* Abingdon: Routledge.

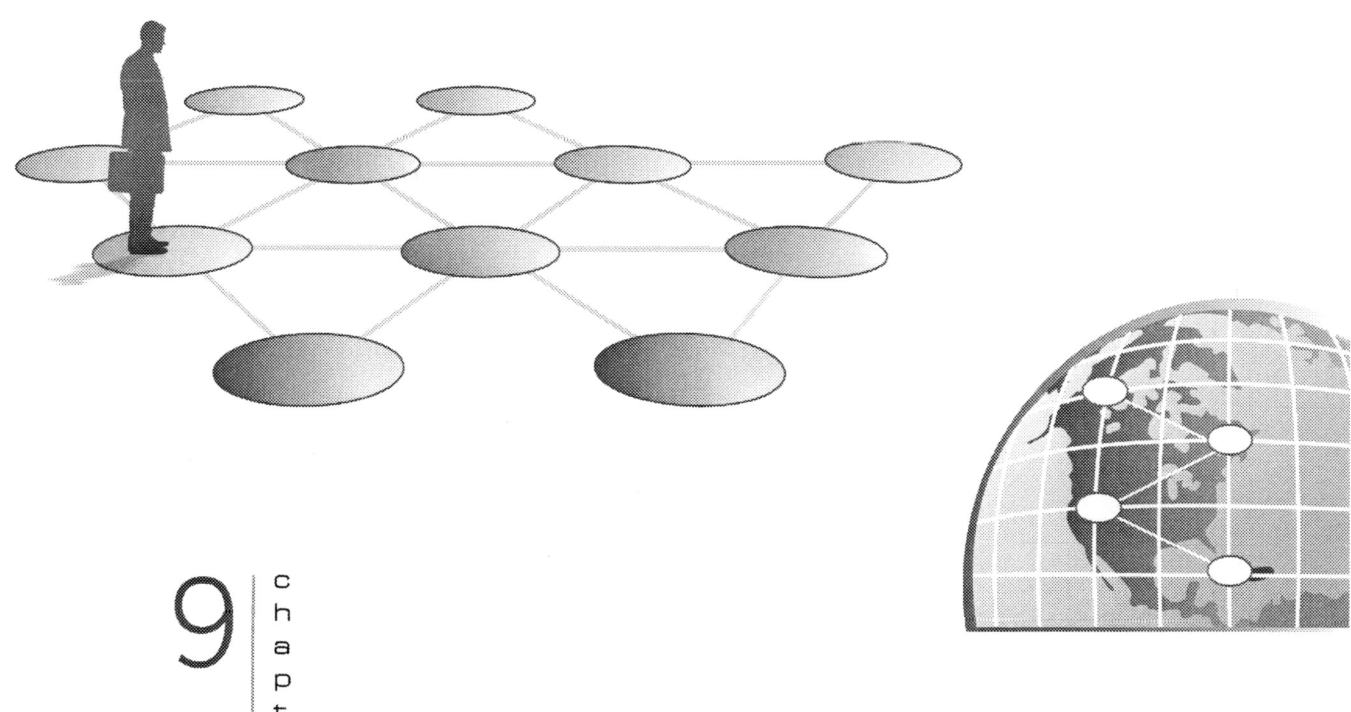

9 | c h a p t e r

The integrated communications plan

In the previous chapter we looked at the need to integrate marketing strategy. Within this chapter we move on to look at the practical stages of developing the plan, managing budgets and resources, the methods used to communicate and finally how to evaluate and measure the plan.

Within this chapter we outline the contents and format of the communications plan itself and also consider implementation issues.

The various frameworks used for communications planning are compared and you will be able to use the framework of your choice within your own organisation.

The importance of metrics in marketing is increasingly growing in significance. These metrics often require significant research work and so although research is outside of the scope of this syllabus, we will briefly touch upon some key methods.

Contents

Chapter learning outcomes

By the end of the chapter you will be able to:

- Enhance your understanding of the concept of integrated plans
- Use a planning framework to develop an integrated communications plan
- Prepare and manage communications budgets
- Plan communications tools in a relevant manner within the integrated plan
- Evaluate and measure the success of the integrated plan

1 Developing an integrated plan

1.1 Integrating communications to assist relationships

Marketers cannot control all the information that customers gather and process about products, but nevertheless marketing organisations must use these principles of communication in order to develop an effective coordinated marketing communications plan, and thereby exert some form of control.

Such plans are vitally important in the modern market, because the marketing mix variables on which marketers have traditionally relied to distinguish themselves from their rivals (product design, lower prices, distribution channels etc) have been changed by the march of technology. Competitors can copy what is done quicker than ever before.

Marketing communications is one area where it is still possible to differentiate your product or service – making the customer believe what you want him to believe about your company, product, brand or service. Communications are therefore essential in supporting customer (indeed all stakeholder) relationships.

- The organisation needs to use **extensive contact**, interaction and feedback to learn about (and from) stakeholders, with the aim of continually adding value. *'In order to leverage relationship marketing, marketers need to move from monologue to dialogue with customers [and other stakeholders].'* (Allen et al, 2001)

- The organisation needs to maintain direct and regular **stakeholder communication**, through multiple points of contact and across a range of reasons for contact, in order to develop relationship ties with stakeholders.

- The organisation needs to maintain **multiple exchanges with a number of stakeholders** (network relationships) rather than a single focus on customers, in order to manage all links in the customer value delivery chain.

- **Dialogue** and **developing trust** provide a basis for the customisation and personalisation of contacts, messages and value-propositions, which further deepen stakeholder relationships.

- Communication is part of the **relationship value offered to stakeholders**: keeping them informed in areas of their interest or concern, guiding and supporting them through changes in the organisation's plans, communicating support for their causes.

Global case study

Toshiba gigabeat

In 2005 to support the product launch of its new gigabeat portable hard drive devices, Toshiba's Marketing Communications department developed an integrated marketing campaign that incorporated various components such as advertising, public relations, online, promotions and point of sale. Based on

Developing Integrated Communications Strategy

the theme of "Music in Color™", the new ad campaign featured well-known artists such as Blues Traveler, Joan Jett and the Blackhearts and the new emerging band, Vendetta Red. To celebrate the launch, Joan Jett and the Blackhearts hosted the gigabeat launch party at the Times Square Studios in New York featuring a live performance by Blues Traveler.

The "Because Music is Just Better in Color™" advertising campaign, ran in major media print and broadcast outlets, which emphasied the product's innovative 2.2 inch display. The campaign featured TV spots appearing on MTV networks' properties such as VH-1®, MTV™, Comedy Central® and TV Land™. In conjunction with the TV campaign, print ads highlighted the "Because Music is Just Better in Color" theme. Toshiba also partnered with Rolling Stone magazine to promote gigabeat both in print and online.

The result was an integrated marketing campaign and strong partnerships that allowed Toshiba to leverage its brand at retail in order to develop a presence in the crowded MP3 market.

Source: adapted from www.toshiba.com [accessed 10.11.09]

The Centre for Integrated Marketing has published many papers, case studies and presentations. A selection can be found on their website at:
www.centreforintegrategmarketing.com

1.2 The process and structure of a communications plan

The nature of the prevailing market conditions is an important context that needs to be considered when planning and developing marketing communications activities.

Like all other areas of business activity, marketing communications needs careful housekeeping. The amount of money available is the key constraint on marketing communications, and there is growing pressure on total communications expenditure. This is because of fluctuating world economies, increasing media costs and also because methods of measuring the effectiveness of spending have been improved, and wastefulness is more transparent.

1.2.1 Communications planning frameworks

Each marketing communications programme is developed in unique circumstances. It is vitally important that the contextual conditions are analysed in order that any factor that may influence the content, timing or the way the audience receives and interprets information, be identified and incorporated within the overall plan.

Fill (2009) and Smith and Taylor (2004) cite the SOSTAC® model as suitable planning framework. We have already covered SOSTAC® in Chapter 5 when we looked at its use for planning digital marketing strategies. We noted at that point that it is a generic framework which outlines six stages of the planning process:

S	Situation	Where are we now? What is the context
O	Objectives	Where do we want to go?
S	Strategy	How do we get there? Who are we going to talk to so we will get there?
T	Tactics	The details of the strategy Communication tools employed What is the exact message?
A	Action	Implementation – putting the plan into operation
C	Control	What measures, monitoring, reviewing and modifying will be used?

Assessment advice

Each of the SOSTAC® (or the alternative framework you chose) elements are expected to be included (with full justification) within a communications plan.

Despite Fill (2009) suggesting that SOSTAC® is an adequate tool for planning he has also criticised the model because it is not specific for communications plans and can be utilised for any planning. The danger being that the focus on communication is diluted at the situation analysis stage. Fill (2009) argues that the tendency to fall into a general marketing plan or SWOT analysis which means that the justification of the communication strategy and promotional mixes is lost. **There is an implicit need to identify and understand the characteristics of the target audience.**

1.2.2 The layout of the communications plan

MARKETING COMMUNICATIONS PLAN FOR [IDENTIFIED CUSTOMER if relevant]		
1	**Communications objectives**	• What problem/need the communication plan is designed to address • What the communications plan is intended to achieve (SMART objectives if possible) • Co-ordinating marketing mix strategies within which the plan has been developed
2	**Target audience**	• Stakeholder group targeted by the plan • Key needs, concerns, interests and drivers of the group • Information needs of the group (either in general or in relation to the specific problem/situation) • Media and communication tools most used by and influential for the stakeholder group
3	**Core message(s)**	• Purpose of the message: desired stakeholder response • Content: key points of the message, and how best conveyed (eg text, multi-media) • Style: informative, persuasive, personal etc. • How the message fits within the co-ordinated marketing mix: consistency, synergy
4	**Communication media and tools**	Which media and tools will be used (with explanation/justification of each in terms of their relevance to the objectives and appropriateness for the target audience): • Advertising • Direct Marketing • Public Relations • Digital tools *And so on...*

5	**Timetable**	• Period over which communication will be required
		• Timescales for review and measurement
6	**Resource allocation**	Estimated expenditure (or basis on which budget should be set: see later section of this chapter)
7	**Monitoring and control**	How progress and results will be monitored, reviewed and measured against objectives (see later section of this chapter)

Assessment advice

Typical problems with the actual document associated with the plan is that there is insufficient detail. To avoid this in your own assignment pay attention to the following:

- Ensure your **objectives are clear** because it is from these that the entire strategy should develop

- Remember to fully **justify** all your recommendations throughout the plan

- **Be specific** particularly when talking about **target markets** and audiences

- Clearly outline your exact schedule of proposed activities, **timetables, Gannt charts, project timelines** are all used to create media schedules and implementation plans for the projects

- Be transparent with regard to **costs and the budget** – provide breakdowns where necessary

- **Break down resources** required and allocate specific action points for key departments

- Don't just outline the metrics, but also discuss the **process for measurement**, when this is to happen and who is responsible.

1.3 Situation analysis

In order to help provide for a systematic appraisal of the prevailing and future conditions, a context analysis is recommended when formulating a marketing communications programme. This consists of a review of the various sub-contexts.

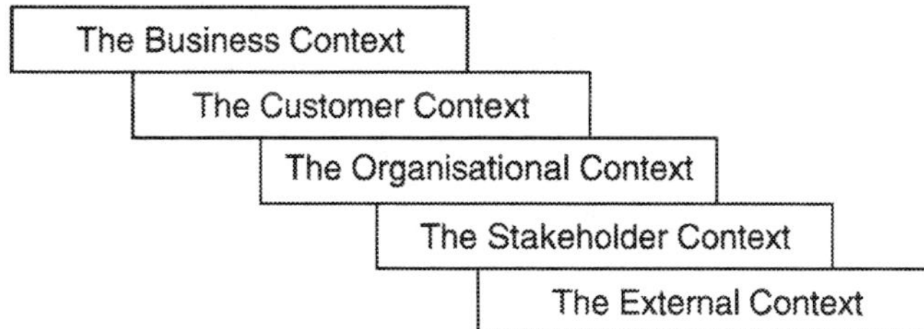

The business context

This part of the analysis involves a consideration of the markets and conditions in which the organisation is operating, which are of prime concern for the coordinated marketing communication programme.

Competitors' communications, general trading conditions and trends, the organisation's corporate and marketing strategies, a detailed analysis of the target segment's characteristics and a brand audit are the primary activities associated with this context.

The customer context

Here the emphasis is upon understanding buyer behaviour and the decision-making processes that buyers in the market exhibit. The objective is to isolate any key factor in the process or any bond that customers might have with the product/brand. This can then be reflected in any communication. Segmentation is

critical at this stage. To be effectively used within a plan Smith (2004) maintain that segments should satisfy the following criteria:

- Be **measurable** or quantifiable
- Be **substantial** to warrant attention
- Be **accessible** so that they can be communicated with
- Be **relevant** – or rather the product or service must be relevant to the segment

The stakeholder context

Coordinated marketing communications recognise that there are audiences other than customers, with whom organisations need to communicate. For example, members of the marketing channel, the media, the financial community, local communities and shareholders all seek a dialogue with the focus organisation. The strength and duration of the dialogue may vary but messages need to be developed and communicated and the responses need to be understood and acted upon wherever necessary.

The organisational context

The characteristics of the organisation can impact heavily on the nature and form of the communications they enter into. It is important, therefore, to consider the culture and the strength of identity the workforce has with the organisation. This is of absolute importance if truly coordinated communications are to be forged. In order to appreciate the strength of this subcontext, think about the way the staff of different companies communicate with you as a customer. Internal and external audiences communicate with each other and this is a significant part of coordinated marketing communications. Industrial segmentation should be applied at this contextual level.

The external context

Coordinated marketing communications are influenced by a number of factors in the wider environment. These political, economic, social and technological elements are largely uncontrollable by organisations, for example economic conditions or laws and regulations. Nevertheless, they can shape and determine what, when and how messages are communicated to audiences.

1.4 Setting marketing communications objectives

In order to deliver an effective plan, it is important to establish marketing communications objectives. These will involve variables such as perception, attitudes, developing knowledge and interest or creating new levels of prompted and spontaneous awareness. As we have seen, they also need to be SMART.

Ultimately, communications are designed to meet three objectives.

Awareness	Increase brand awareness and establish brand recognition
Trial	Stimulate trial purchase
Reinforcement	Stimulate and reinforce brand loyalty

1.4.1 Sales goals

If you ask most people what the goal of marketing communications is, then most will respond 'to increase sales'. Ultimately this (and profit) is an important outcome, but ask yourself this: are sales generated by marketing communications alone? What role does each of the other elements of the marketing mix play? How will sales vary if a competitor reduces its prices or you increase yours? What impact do marketing channel and product availability play in sales performance? Marketing communications is important but it is not the sole contributor to marketing success or failure.

Sales goals are important and performance can be determined in terms of sales volumes, sales value or revenue, market share, or profitability measures such as return on investment (ROI). They are a useful management aid as they are easy to comprehend and measurement is straightforward.

1.4.2 Communication based goals

There are many situations when the aim of a communication campaign is to enhance the image or reputation of an organisation or product (Fill, 2009). Sales in this instance are not regarded as the sole goal. Several models have been used to encourage the efficient and quantified use of promotional objectives. We looked at Hierarchy of Effects models in Chapter 1 which are used for this purpose. AIDA and DAGMAR models are frequently used to discuss the various stages a buyer goes through prior to purchase. These can be linked to the promotional objectives.

AIDA stands for:

- Attention
- Interest
- Desire
- Action

DAGMAR stands for Defining Advertising Goals for Measured Advertising Results. There are a number of stages to communications messages:

- **Awareness** – products and services have to exist for consumers otherwise they will not be aware they can make a purchase

- **Comprehension** – audiences need information and knowledge about the product attributes

- **Conviction** – audiences need to be persuaded that the product is superior to alternatives

- **Action** – potential buyers need to be encouraged and helped to make a purchase, sales people can help consumers act upon convictions, calls to action within messages include adding a website URL, ease of reply, contact details etc

1.4.3 Example objectives

The following are examples of typical communication based objectives:

- To increase awareness from 40% to 50% within twelve weeks of the campaign launch among 35-45 year old 'sporty' women.

- To position the spa as the friendliest luxury health spa within the Asian region within a one year period.

- To support the launch of the Global Marketer programme by generating 20% awareness amongst current global marketing students within six months.

1.4.4 Identifying and gaining new prospects as an objective

In order to achieve a sale, each buyer must move, or be moved, through a series of steps. These steps are essentially communication-based stages whereby individuals learn more about a product and mentally become more disposed towards adjusting their behaviour in favour (or not) of purchasing the item (as described in the AIDA and DAGMAR models above).

Awareness is an important state to be achieved as without awareness of a product's existence it is unlikely that a sale is going to be achieved! To achieve awareness people need to see or perceive the product, they need to understand or comprehend what it might do for them (benefits) and they need to be convinced that such a purchase would be in their best interest and to do this there is a need to develop suitable attitudes and intentions.

Analysis of the organisational context will have determined the extent to which action is required to communicate with members and non-members. Corporate communications, particularly with employees, and corporate branding to develop the image held by key stakeholders, should be integral to such coordinated marketing communication campaigns. These tasks form a discrete part of the communication programme.

In addition to this, it is the responsibility of the communication programme to communicate the mission and purpose of the organisation in a consistent and understandable form. And, the organisation needs to be able to listen and respond to communications from their stakeholders in order that they are able to adjust their position in the environment and continue to pursue their corporate goals. The strategy for communications will therefore need to take into account a variety of contexts.

1.5 Communications strategy in different contexts

An organisation's interaction with the various markets in which it operates is, of course, crucial. In order that its marketing communications be effective it is necessary to understand the conditions and elements that prevail in specific markets.

- Is the market expanding or contracting?
- What are the values and beliefs held by the target audience towards the firm's products and those of competitors?
- What are the attitudes of intermediaries?
- What is the nature of competitive communications?

Most of our discussion so far has tended to concentrate upon the marketing of goods by businesses to consumers (B2C). Although the principles of marketing communications are the same for both consumer and industrial markets, there are significant differences in the details of how promotion is carried out. In particular, the targets in industrial markets are usually more specific and promotional budgets are usually more limited.

Perhaps the most significant differences are the nature of the buying motivation and the linked nature of the buying decision process. In industrial buying there are many motivations. These stem partly from the technical use of the product but also from financial, security of supply and, to a lesser degree, emotional reasons.

Decision makers	Buying motivation
Operations Manager	Uses the product in the organisation's processes – wants efficiency and effectiveness
Technical Manager	Often has to test and approve the product – wants reliability
Managing Director	May approve major expenditure or change of supplier
Purchasing Manager	Approves conditions of purchase Monitors supplier performance
Legal Manager	Draws up or approves legal contracts with supplier
Finance Manager	Approves expenditure and controls debt payment
Health and Safety Manager	May have a role to play with hazardous supplies

It will be obvious that marketing communications strategy for industrial marketing must reflect this considerably more complex decision-making process.

The variety of products in business markets is extremely large. Business products vary from product inputs to items for resale. They can be broken down into three main types.

- Capital equipment (major purchases of fixed assets)
- Production inputs (becoming part of the buyer's process)
- Business supplies/services (ongoing use by the buyer)

Again, each type of purchase will need a different communications strategy. Later in the chapter we focus on specific promotional tactics, in the meantime, the range of promotional methods is described below.

Method	Comment
Personal selling	This is a major component of industrial marketing because of the need to deal with technical and other issues on a face to face basis.
Internal selling	Increasingly it is recognised that a salesperson has an internal role to play in representing his customers' needs to the company.
Internet	The use of the Internet for e-commerce is perhaps more highly developed in industrial marketing than in the consumer sector. Advertising and online catalogues are just two of the ways that it can be used. Many companies have also set up electronic links with suppliers and customers for such functions as automatic ordering.
Advertising	A wide variety of publications exist which can be used to target individual market sectors including: (i) trade journals (ii) business press (iii) directories
Telemarketing	Telemarketing has been proved to be a very cost effective method of order processing, customer service, sales support and account management.
Direct mailing	Direct mail, another form of direct marketing, has been used by industrial marketers for a long time but its use has substantially increased. It can be used to provide information and generate enquiries. It can be tailored to individual customer needs.
Public relations	Sometimes in industrial markets this is referred to as publicity. It often focuses on getting editorial coverage in appropriate magazines but it has a wider role of building customer relations.
Sales promotion	Sales promotion is an important area of communication in industrial markets. There are a wide range of methods that are of well established use in industrial campaigns. • Literature • Exhibitions • Videos • Discounting • Events • Business gifts • Trade shows

1.5.1 Influences on the effective audience

The effective audience for a medium, and therefore the competitiveness of different media, is influenced by the following factors.

(a)   **Opportunity to use the medium**. The potential audience will not be able to use TV during working hours, or magazines while driving, or cinema over breakfast. Radio in the morning and TV in the evening have bigger effective audiences.

(b)   **Effort required to use the medium**. People usually use the medium that will cost them least effort. Print media require the ability to read and concentrate: television is comparatively effortless.

(c)   **Familiarity with the medium**. People consume media with which they are familiar: hence the survival of print media, since the education system is still predominantly print-orientated. Electronic media are however gaining ground.

(d)   **Segmentation by the medium**. The print media currently have the greatest capacity for segmentation into special-interest audiences. Commercial television segments to a limited extent through programming, and cable/satellite television to a greater extent, through the proliferation of channels. Some media only charge in proportion to the segment you are targeting, which is more cost effective than paying for the full circulation. (Pickton and Broderick, 2004).

Activity 1

What opportunity, effort and familiarity issues might you consider when appraising the following media?

(a)   A newly launched radio station
(b)   Daytime television
(c)   Posters on buses
(d)   Web pages

In order to be able to incorporate this strategy, a budget and resource plan is required.

2 Budget and resource management

The **principal budget factor** should be identified at the beginning of the budgetary process. It is often sales volume and so the sales budget has to be produced before all the others.

A **budget** is a consolidated statement of the resources required to achieve objectives or to implement planned activities. It is a planning and control tool relevant to all aspects of management activities.

2.1 Purposes of a budget

- **Co-ordinates** the activities of all the different departments of an organisation; in addition, through participation by employees in preparing a budget, it may be possible to motivate them to raise their targets and standards and to achieve better results.

- **Communicates** the policies and targets to every manager in the organisation responsible for carrying out a part of that plan.

- **Control** by having a plan against which actual results can be progressively compared.

2.2 Preparing budgets

Procedures for preparing the budget are contained in the **budget manual**, which indicates:

- People responsible for preparing budgets
- The order in which they must be prepared
- Deadlines for preparation
- Standard forms

The preparation and administration of budgets is usually the responsibility of a **budget committee**. Every part of the organisation should be represented on the committee.

The preparation of a budget may take weeks or months, and the budget committee may meet several times before the master budget is finally agreed. Functional budgets and cost centre budgets prepared in draft may need to be amended many times over as a consequence of discussions between departments, changes in market conditions, reversals of decisions by management, etc during the course of budget preparation.

2.2.1 The budget period

A budget does not necessarily have to be restricted to a one-year planning horizon. The factors which should influence the **budget period** are as follows.

- **Times**. A plan decided upon now might need a **considerable time** to be put into operation. Many companies expect growth in market share to take a number of years.

- **In the short-term some resources are fixed**. The fixed nature of these resources, and the length of time which must elapse before they become variable, might therefore determine the planning horizon for budgeting.

- All budgets involve some element of **forecasting and even guesswork**, since future events cannot be quantified with accuracy.

- Since **unforeseen events** cannot be planned for, it would be a waste of time to plan in detail too far ahead.

- Most budgets are prepared over a one-year period to enable managers to plan and control **financial results for the purposes of the annual accounts**.

2.2.2 The principal budget factor

The first task in budgeting is to identify the principal (key, limiting) budget factor. This is the factor which puts constraints on growth. The principal budget factor could be:

- Normally, sales demand, ie a company is restricted from making and selling more of its products because there would be no sales demand for the increased output at a price which would be acceptable/profitable to the company.

- Resources machine capacity, distribution and selling resources, the availability of key raw materials or the availability of cash.

- Once this factor is defined then the rest of the budget can be prepared.

2.3 Budgets as a control device

> **Definition**
>
> A **budget** is a plan representing the **resources** required to achieve objectives. There are various methods of setting the marketing budget, including the **objective and task** measures.
>
> A **budget** is a consolidated statement of the resources required to achieve desired objectives, or to implement planned activities. It is a planning and control tool relevant to all aspects of management activities.
>
> A **forecast** is an estimate of what might happen in the future.

2.3.1 Problems in constructing budgets

(a) **Unpredictability** in economic conditions or prices of inputs.

(b) Because of **inflation**, it might be difficult to estimate future price levels for materials, expenses, wages and salaries.

(c) **Managers might be reluctant to budget accurately.**

    (i) **Slack**. They may overstate their expected expenditure so that by having a budget which is larger than necessary, they will be unlikely to overspend the budget allowance. (They will then not be held accountable in control reports for excess spending.)

(ii)   They may **compete** with other departments for the available resources, by trying to expand their budgeted expenditure. Budget planning might well intensify inter-departmental rivalry and the problems of 'empire building'.

(d)   **Inter-departmental rivalries** might ruin efforts towards co-ordination in a budget.

(e)   Employees might resist budget plans either because the plans are not properly communicated to them, or because they feel that the budget puts them under pressure from senior management to achieve better results.

2.3.2 Setting the sales budget

(a)   A **preliminary sales estimate** uses the following data.

(i)   A study of normal business growth
(ii)   A forecast of general business conditions
(iii)   A knowledge of potential markets for each product
(iv)   The practical judgement of sales and management staff
(v)   A realisation of the effect on sales of basic changes in company policy

(b)   An **adjustment of the above preliminary sales estimate** may be required

(i)   Seasonal nature of the business
(ii)   Overall production or purchasing capacity
(iii)   Overall selling expenses and net profits
(iv)   The financial capacity of the business

(c)   The adjusted anticipated sales by value and quantity contained in the sales budget should then be classified by commodities, departments, customers, salesmen, countries, terms of sale, methods of sale, methods of delivery and urgency of delivery.

2.3.3 The expense budgets related to marketing

(a)   *Selling expenses budget*

(i)   Salaries and commission
(ii)   Materials, literature, samples
(iii)   Travelling (car cost, petrol, insurance) and entertaining
(iv)   Staff recruitment and selection and training
(v)   Telephones and telegrams, postage
(vi)   After sales service
(vii)   Royalties/patents
(viii) Office rent and rates, lighting, heating
(ix)   Office equipment
(x)   Credit costs, bad debts

(b)   *Advertising budget*

(i)   Trade journal – space
(ii)   Prestige media – space
(iii)   PR space (costs of releases, entertainment)
(iv)   Blocks and artwork
(v)   Advertising agents commission
(vi)   Staff salaries, office costs
(vii)   Posters
(viii) Cinema
(ix)   TV
(x)   Signs

Developing Integrated Communications Strategy

(c)   *Sales promotion budget*

    (i)   Exhibitions: space, equipment, staff, transport, hotels, bar
    (ii)   Literature: leaflets, catalogues
    (iii)   Samples/working models
    (iv)   Point of sale display, window or showroom displays
    (v)   Special offers
    (vi)   Direct mail shots – enclose, postage, design costs

(d)   *Research and development budget*

    (i)   Market research – design and development and analysis costs
    (ii)   Packaging and product research – departmental costs, material, equipment
    (iii)   Pure research – departmental costs materials, equipment
    (iv)   Sales analysis and research
    (v)   Economic surveys
    (vi)   Product planning
    (vii) Patents

(e)   *Distribution budget*

    (i)   Warehouse/deposits – rent, rates, lighting, heating
    (ii)   Transport – capital costs
    (iii)   Fuel – running costs
    (iv)   Warehouse/depot and transport staff wages
    (v)   Packing (as opposed to packaging)

2.3.4 Methods of setting the marketing budget

Method	Comment
Competitive parity	Fixing promotional expenditure in relation to the expenditure incurred by competitors (This is unsatisfactory because it presupposes that the competitor's decision must be a good one.)
The task method (or objective and task method)	The marketing task for the organisation is set and a promotional budget is prepared which will help to ensure that this objective is achieved. A problem occurs if the objective is achieved only by paying out more on promotion than the extra profits obtained would justify.
Communication stage models	These are based on the idea that the link between promotion and sales cannot be measured directly, but can be measured by means of intermediate stages (for example increase in awareness, comprehension, and then intention to buy).
All you can afford	Crude and unscientific, but commonly used. The firm simply takes a view on what it thinks it can afford to spend on promotion given that it would like to spend as much as it can.
Investment	The advertising and promotions budget can be designed around the amount felt necessary to maintain a certain brand value.
Rule-of-thumb, non-scientific methods	There include setting expenditure at a certain percentage of sales or profits.

3 Communication methods

Targeting the communications mix for stakeholders means identifying and using the appropriate communication media and the appropriate modes of communication. This is a critical element within a communications plan which must be fully justified.

There are several decisions facing those responsible for planning marketing communications:

- Who should receive messages?
- What the messages should say?
- What image of the organisation/brand receivers should retain?
- How much is to be invested in the process (as described previously in the budgeting process)?
- How the messages are to be delivered?
- What actions receivers should take?
- How the whole process should be controlled?
- How to determine what was achieved?

Assessment advice

While planning your communications plan for the assignment, use the list above as a checklist to ensure you have covered all that you need to.

It is important to note that effective communication is not a wholly 'one-way' flow. Producers need to understand customer motivations if they are to be able to 'talk' to them in an effective way and properly understand their market.

3.1 Media-Neutral Planning

Media-neutral planning (MNP) emerged as a response to IMC in terms of suggesting that communications should be planned according to the needs of the message and that there should not be a preference shown for one or more, media from the outset.

Traditionally agencies have been rewarded with higher fees for designing mass media campaigns. There became a culture of preference for mass media recommendations as a result. MNP advocates however that the most suitable method for the message should be selected which highlights the need for possibly different agency remuneration negotiations or working practices.

Definition

Media-neutral planning attempts to stimulate the use of a communications mix that is not mass media-orientated. This means that rather than keep recommending that clients use mass media communications a more balanced mix of tools and media are adopted to be more efficient and effective (Fill, 2009).

3.2 The promotional mix

From any earlier studies in marketing communication, you should be familiar with the various communication tools available to the marketer. To remind you, we will cover some of the tools but not is significant detail.

The **promotion** mix is the total marketing communications programme of the organisation, consisting of a specific combination or blend of promotional tools used to reach the target audience for a given marketing task. The full range of tools that can be used to secure favourable responses from, and build sustainable relationships with, stakeholder audiences is shown in the following diagram.

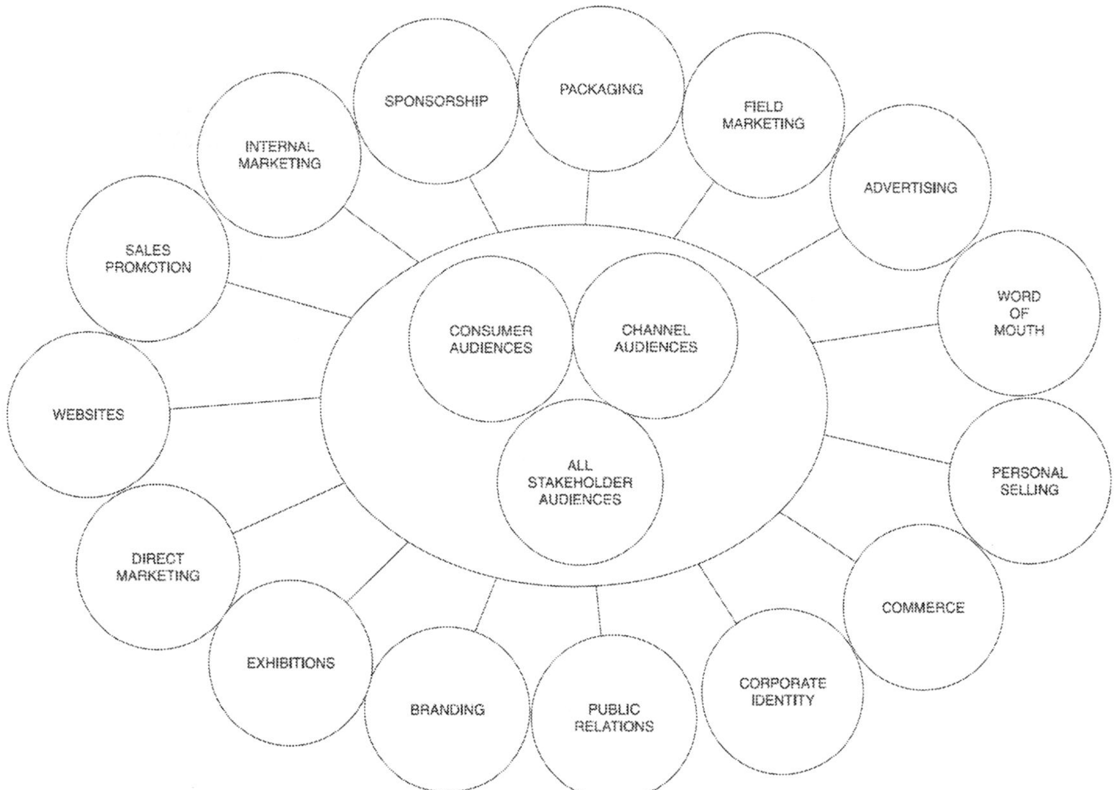

If the sheer range of tools available seems a bit intimidating, we have bad news – and good news:

- **The range of promotional tools and media continues to grow.** The variety of media has increased (or been fragmented) in many ways. There are more print media (publications aimed at more and more highly defined niche segments) and more broadcast media (with developments in satellite, cable and digital TV, DVD, Webcasting and Podcasting and so on). Marketing messages are being put on more and more surfaces, from buildings to tabletops to Post-It Notes – and even people!

- **There is no 'one best mix' for any given message in any given market.** While some communication tools may be identified as more or less effective in different contexts, selecting and combining promotional tools is still very much an art – not a science!

Activity 2

Which of the tools in diagram does your work organisation, or other organisation that you are studying, use? What audience(s) is each tool designed to address?

Which tools are particularly used to address shareholder, supplier, intermediary, pressure group and wider community audiences?

3.3 Legal and ethical aspects of marketing communications

Laws and regulations governing marketing communications must obviously be observed. Each country will have its own set of restrictions which apply to advertising, packaging, sales promotion or direct marketing.

In the EU alone, there are a number of significant differences regarding the regulation of advertising between member states. A Gossard TV commercial came under scrutiny in the UK for its risqué execution. In France, the problem was not the generous display of cleavage, but the fact that the advert was set in a bar where alcohol was being consumed. There is a ban on TV alcohol adverts and the Gossard ad needed to be re-edited to fall in line with French restrictions.

In some countries, restrictions apply to the use of non native models and actors. This can mean that advertising has to be reshot for specific countries.

Packaging regulations can vary. In a number of European markets, the push towards environmentally-friendly packaging has resulted in far more stringent rules than apply in the UK. In Denmark, soft drinks may not be sold in cans, only in glass bottles with refundable deposits.

Sweden forbids all advertising aimed at children under 10; Greece bans TV toy advertising between 7am and 10pm; some countries require ads for sweets to carry a toothbrush symbol; others have rules intended to curb advertisers from encouraging children to exercise 'pester power'.

The same maze of national rules exists when it comes to promoting alcohol, tobacco, pharmaceuticals and financial services. Price advertising and discounting measures are so disparate that cross-border campaigns using discounts are all but impossible... In Germany, cash discounts are limited to 3% and the advertising of special offers is also restricted. Austria, Belgium and Italy also have strict regimes. In contrast, in Scandinavia, where the advertising law is more closely linked to consumer protection rather than unfair competition considerations, price advertising is encouraged – Swedish law, for example, promotes comparative price advertising between traders.

Critics of marketing argue that it is dedicated to selling products which are potentially damaging to the health and well-being of the individual or society. Examples include tobacco, alcohol, automobiles, detergents and even electronic goods such as computers and video recorders. It has been argued that even seemingly beneficial, or at least harmless, products, such as soft drinks, sunglasses or agricultural fertiliser, can damage individuals and societies. In traditional societies, new products can disrupt social order by introducing new aspirations, or changing a long established way of life.

3.3.1 Ethics and the law

Ethics deal with personal moral principles and values, but laws are the rules that can actually be enforced in court. However, behaviour which is not subject to legal penalties may still be unethical. We can classify marketing decisions according to ethics and legality in four different ways.

- Ethical and legal (eg the Body Shop)

- Unethical and legal (eg 'gazumping')

- Ethical but illegal (eg publishing stolen but revealing documents about government mismanagement)

- Unethical and illegal (eg employing child labour)

Different cultures view marketing practices differently. While the idea of intellectual property is widely accepted in Europe and the USA, in other parts of the world ethical standards are quite different. Unauthorised use of copyrights, trademarks and patents is widespread in countries such as Taiwan, Mexico and Korea. According to a US trade official, the Korean view is that ' ...*the thoughts of one man should benefit all*', and this general value means that, in spite of legal formalities, few infringements of copyright are punished.

3.3.2 Ethics in marketing communications

Ethical considerations are particularly relevant to promotional practices. Advertising and personal selling are areas in which the temptation to select, exaggerate, slant, conceal, distort and falsify information is potentially very great indeed.

Questionable practices here are likely to create cynicism in the customer and ultimately to preclude any degree of trust or respect for the supplier. Even taking into account the protection afforded to the consumer by different acts of legislation throughout the world, many think that persuading people to buy something they don't really want is intrinsically unethical, especially if hard sell tactics are used.

Also relevant to this area is the problem of corrupt selling practices. It is widely accepted that a small gift such as a mouse mat or a diary is a useful way of keeping a supplier's name in front of an industrial purchaser. Most business people would however condemn the payment of substantial bribes to purchasing officers to induce them to favour a particular supplier.

But where does the dividing line lie between these two extremes?

Assessment advice

Research the laws relating to marketing communications and general customs relating to 'gifts' within your own county. If you work for a multinational corporation, how are the variances in legal acts dealt with?

3.3.3 Consumer and community issues

The consumer movement can be defined as a collection of organisations, pressure groups and individuals who seek to protect and extend the rights of consumers. The movement originated in a realisation that the increasing sophistication of products meant that the individual's own judgement was no longer adequate to defend against inappropriate marketing.

3.4 Why advertise

(a) To **promote sales**

Advertising is particularly good at raising awareness, informing and persuading. It can be used to stimulate primary demand for a product (eg in the introduction of a new product) and selective demand for a particular brand (eg in competition with other brands). This works in intermediary markets (eg selling in to retailers) as well as consumer markets, effectively 'introducing' the product in advance of a sales call. One example of a different sort of response is recruitment advertising: promoting the organisation and the job within the recruitment market, in order to secure applications from the best potential recruits. The same kind of approach may be used to advertise for suppliers (eg by putting a contract out to tender).

(b) To **create an image**

Institutional advertising is used by companies to improve their public image, and by not-for-profit and public sector organisations to promote their programmes (eg persuading people not to drink and drive, to support a pressure group or donate to a charity). Marketing organisations often use advertising to promote their corporate social responsibility credentials to the wider stakeholder audience. Innocent Drinks, the subject of the sample case study, is an example of such an approach.

(c) To **support sales staff**

Advertising can support personal selling, for example, by raising customer (consumer or intermediary) awareness of the product/service, motivating them to contact sales representatives. The company's sales promotions, presence at exhibitions, website and other communication channels can also be brought to the target audience's attention by advertising.

(d)   To **offset competitor advertising**

Companies often attempt to defend their market share by responding aggressively to competitors' advertising campaigns. Advertisements may also be used to counter negative public relations messages and alter public opinion (eg correcting negative impressions given by critical incidents such as a product recall, or countering negative pressure group statements).

(e)   To **remind and reassure**

Advertising reinforces the purchase decision and repeat purchase, by reminding consumers that the product continues to be available – and offers benefits – and reassuring them that they have made the right choice. In industrial markets, advertising may add credibility to sales visits by demonstrating professionalism and expenditure.

3.5 Developing an advertising strategy

Jobber (2007) suggests the following framework for developing an advertising strategy.

As indicated by the diagram, the planning of an advertising campaign includes:

(a)   **The identification of the target audience**: Who are they? Where are they? Which demographic group do they fall into? What are their interests, media consumption habits, buying patterns, attitudes and values?

(b)   **The specification of the communication or promotional message**: What do you need to say, and in what way, in order to impact on the audience is such a way as to achieve your marketing objectives?

(c)   **The setting of targets**: What is the marketing goal to which the advertising can contribute? What do you expect the ad to achieve, and at what cost? What aspects of the audience's thinking or behaviour do you wish to change, and how will you recognise and measure that change if and when it occurs?

When it comes to media decisions, the general criteria for selecting the appropriate medium to convey the promotional message to the audience are as follows:

- The advertiser's specific objectives and plans
- The size of the audience that regularly uses the medium
- The type of people who form the audience for the medium
- The suitability of the medium for the message
- The cost of the medium in relation to its ability to fulfil the advertiser's objectives
- The susceptibility of the medium to testing and measurement

Activity 3

All other things being equal, and subject to detailed research, at what time of day, or in what kind of programme, might you advertise the following products on TV?

(a)   Shoe polish

(b)   Home disinfectant

(c)   Car repairs

(d)   Chat/introduction lines

3.5.1 The size and type of the audience

Each medium reaches a certain number and 'type' (demographic group, market segment, interest group) of people. There is a trade-off between the size and relevance of the available audience.

(a)   General interest, national mass-market media (such as national newspapers or TV) will have the highest circulation figures, but may not reach the highest percentage of the target market segment.

(b)   Segmentation may be possible through the scheduling and placing of ads in large-scale media (for example, in special-interest sections of supplements in the press, or by programme preference on TV).

(c)   Targeted media may reach a smaller population, but a higher percentage of the target audience.

- Local or regional media
- Specialist magazines and journals
- Media which fit with the 'media habits' of the target audience

3.6 Direct response advertising

> **Definition**
>
> **Direct response advertising** appears in the prime media (TV, press etc) and is designed to elicit 'an immediate response from the consumer in terms of purchase, or request for a brochure, or a visit to the shop.' (Blythe, J., 2005 p. 250). This may include a freephone telephone number, or a website address. It does not involve the use of intermediaries.

Direct response advertising may involve traditional advertising in a newspaper or magazine with a cut out (or stuck on) response coupon; loose inserts with response coupons or reply cards; direct-response TV or radio advertisements, giving a call centre number or website address to contact.

Advertising on interactive TV includes a 'pop up' button which gives you the option to interact by transferring to a website. DRTV (home shopping) is presently conducted mainly through the use of television commercials or infomercials (which combine information with a commercial), which direct customers to a website or telephone order lines. In the UK, cable and satellite also provide a number of channels exclusively devoted to shopping.

Direct response advertising enables detailed measurement of the effectiveness of ads on different stations, at different times, in different formats.

Direct mail tends to be the main medium of direct response advertising. The main reason for this is that other major media such as newspapers and magazines are familiar to people for advertising in other contexts. Direct mail has a number of strengths as a direct response medium.

(a)   The advertiser can target down to **individual level**.

(b)   The communication can be **personalised**. Known data about the individual can be used, while modern printing techniques mean that parts of a letter can be altered to accommodate this.

(c)   The medium is good for **reinforcing interest** stimulated by other media such as TV. It can supply the response mechanism (a coupon) which is not yet available in that medium.

(d)   The opportunity to use **different creative formats** is almost unlimited.

(e)   **Testing potential is sophisticated**: a limited number of items can be sent out to a 'test' cell and the results can be evaluated. As success is achieved, so the mailing campaign can be rolled out.

3.7 Personal selling

Personal selling encompasses a wide variety of tasks including prospecting, information gathering and communicating as well as actually selling.

Jobber (2007, p.554) identifies the following elements in the modern selling role:

- Customer relationship management
- Database and knowledge management
- Marketing the product
- Problem solving
- Satisfying needs and adding value

Definition

Personal selling is the presentation of products and persuasive communication to potential clients by sales staff employed by the supplying organisation. It is the most direct and longest established means of promotion within the promotional mix.

Personal selling, or sales force activity, must be undertaken within the context of the organisation's overall marketing strategy. For example, if the organisation pursues a '**pull**' strategy, relying on massive consumer advertising to draw customers to ask for the brands, then the role of the sales force may primarily be servicing, ensuring that retailers carry sufficient stock, allocate adequate shelf space for display and co-operate in sales promotion programmes.

Conversely, with a '**push**' strategy, the organisation will rely primarily on the sales force to persuade marketing intermediaries to buy the product.

3.7.1 Sales roles

A salesperson might perform any of six different activities.

Activity	Salesperson's role
Prospecting	Gathering additional potential customers
Communicating	Communicating information to existing and potential customers about the company's products and services can take up a major proportion of the salesperson's time
Selling	Approaching the customer, presenting benefits, answering objections and closing the sale
Servicing	Providing services to the customer, such as technical assistance, arranging finance and speeding delivery
Information gathering	Feedback and marketing intelligence gathering
Allocating	Allocating products to priority customers, in times of product shortages

3.7.2 The selling process

Personal selling is part of an integrated promotional strategy. It will be supported by a range of other activities such as advertising and PR, lead generation and sales support information.

Elements of the selling process can be depicted as follows.

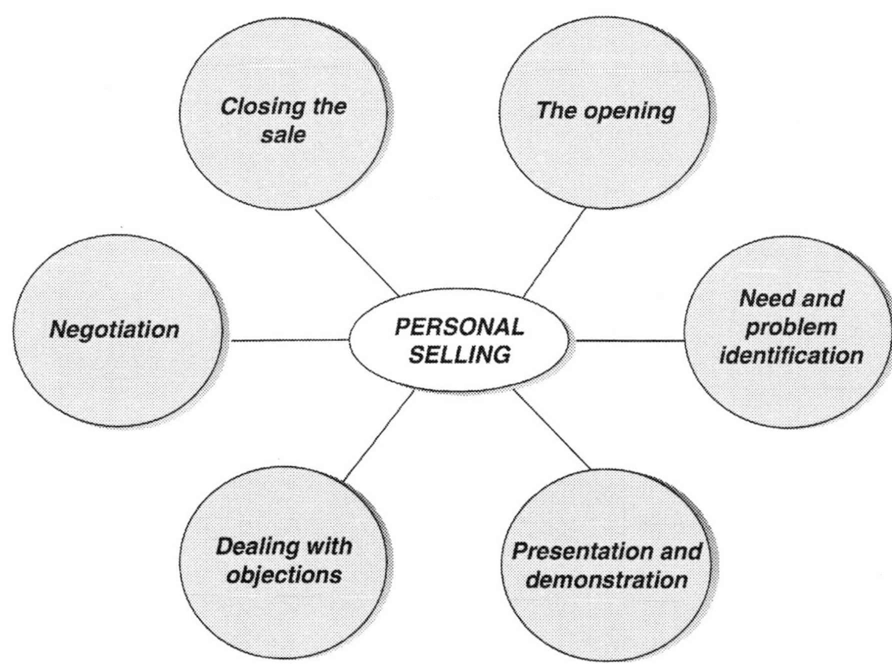

Elements of personal selling

The stages need not occur in any particular order. Objections may occur during the presentation; negotiation may begin during problem identification; and if the process of selling is going well, the salesperson may try to close the sale.

The salesperson's job begins before meeting the buyer. **Preparation** could include finding out about the buyer's personal characteristics, the history of the trading relationship, and the specific requirements of the buyer and how the product being sold meets those requirements. In this way, the salesperson can frame sales presentations and answers to objections.

At the other end, the selling process does not finish when the sale is made. Indeed, the sale itself may only be the start of a long-term **relationship** between buyer and seller.

Personal selling will be **supported** by a range of other marketing communication activities.

(a)   **Product advertising, public relations and sales promotion**, drawing consumer attention and interest to the product and its sources *and* motivating distributors/retailers to stock and sell the product.

(b)   **'Leads'** (interested prospective customers) generated by contacts and enquiries made through exhibitions, promotional competitions, enquiry coupons in advertising and other methods.

(c)   **Informational tools** such as brochures and presentation kits. These can add interest and variety to sales presentations, and leave customers with helpful reminders and information.

(d)   **Sales support information**: customer/segment profiling; competitor intelligence; access to customer contact/transaction histories and product availability. (This is an important aspect of customer relationship management, enabling field sales teams to facilitate immediate response and transactions without time-lags to obtain information.)

3.7.3 The advantages and disadvantages of personal selling

Personal selling is often appropriate in **B2B markets**, where there are fewer, higher-value customers who are looking for a more complex total offering tailored to a more specific set of requirements.

A number of advantages are associated with using personal selling when compared to other promotional tools.

(a) Personal selling contributes to a relatively **high level of customer attention** since, in face-to-face situations, it is difficult for a potential buyer to avoid a salesperson's message.

(b) Personal selling enables the salesperson to **customise the message** to the customer's specific interests and needs.

(c) The two-way communication nature of personal selling allows **immediate feedback** from the customer so that the effectiveness of the message can be ascertained.

(d) Personal selling communicates a larger amount of **technical and complex information** than would be possible using other promotional methods.

(e) In personal selling there is a greater ability to **demonstrate** a product's functioning and performance characteristics.

(f) Frequent interaction with the customer gives great scope for the **development of long-term relations** between buyer and seller.

The main disadvantage of personal selling is the **cost** inherent in maintaining a salesforce. In addition, a salesperson can only interact with one buyer at a time. However, the message is generally communicated more effectively in the one-to-one sales interview, so the organisation must make a value judgement between the effectiveness of getting the message across against the relative expense.

According to Jobber (2007) there are forces at work that are changing the face of selling in the 21st century.

- Rising customer expectations
- Customer avoidance of traditional 'buyer versus seller' negotiations
- Expanding power of major buyers
- Globalisation of markets
- Fragmentation of markets
- Technological forces

3.8 Direct marketing

> **Definition**
>
> **Direct marketing** creates and develops a direct relationship between the consumer and the company on an individual basis.
>
> (a) The Institute of Direct Marketing in the UK defines direct marketing as 'The planned recording, analysis and tracking of customer behaviour to develop relational marketing strategies'.
>
> (b) The Direct Marketing Association in the US defines direct marketing as 'An interactive system of marketing which uses one or more advertising media to effect a measurable response and/or transaction at any location'.

Although it originated in direct mail and mail order catalogues, direct marketing today involves use of a wide variety of media to communicate directly with the target market and to elicit a measurable response. Such media include telemarketing, DRTV, the Internet and online shopping.

3.8.1 Features of direct marketing

It is worth studying these definitions and noting some key words and phases.

(a) **Response**. Direct marketing is about getting people to respond by post, telephone, e-mail or web form to invitations and offers.

(b) **Interactive**. The process is two-way, involving the supplier and the customer.

(c)   **Relationship**. Direct marketing is in many instances part of an on-going process of communicating with and selling to the same customer.

(d)   **Recording and analysis**. Response data are collected and analysed so that the most cost-effective procedures may be arrived at.

Direct marketing helps create and develop **direct one-to-one relationships** between the company and each of its prospects and customers. This is a form of **direct supply**, because it removes all channel intermediaries apart from the advertising medium and the delivery medium: there are no resellers. This allows the company to retain control over where and how its products are promoted, and to reach and develop business contacts efficiently.

3.8.2 Tools of direct marketing

No longer synonymous with 'junk mail', direct marketing tools include direct mail, email, text message, DRTV advertising and telemarketing.

Direct marketing is the fastest growing sector of promotional activity. It now embraces a range of techniques, some traditional – and some based upon new technologies.

(a)   **Direct mail (DM)**: a personally addressed 'written offering' (letter and/or sales literature) with some form of response mechanism, sent to existing customers from an in-house database or mailing list.

Global case study

Computers now have the capacity to operate in three new ways which will enable businesses to operate in a totally different dimension.

Customers can be tracked individually. Thousands of pieces of information about each of millions of customers can be stored and accessed economically.

Companies and customers can interact through, for example, phones, mail, email and interactive kiosks. For the first time since the invention of mass marketing, companies will be hearing from individual customers in a cost-efficient manner.

Computers allow companies to match their production processes to what they learn from their individual customers – a process known as 'mass customisation' which can be seen as 'the cost-efficient mass production of products and services in lot sizes of one'.

There are many examples of companies which are already employing or experimenting with these ideas. In the US Levi Strauss, the jeans company, takes measurements and preferences from female customers to produce exact-fitting garments. The approach offers the company tremendous opportunities for building learning relationships.

The Ritz-Carlton hotel chain has trained staff throughout the organisation to jot down customer details at every opportunity on a guest preference pad. The result could be the following: You stay at the Ritz-Carlton in Cancun, Mexico, call room service for dinner, and request an ice cube in your glass of white wine. Months later, when you stay at the Ritz-Carlton in Naples, Florida, and order a glass of white wine from room service, you will almost certainly be asked if you would like an ice cube in it.

(b)   **Email**: messages sent via the Internet from an email database of customers. Emails can offer routine information, updates and information about new products. Email addresses can be gathered together via enquiries and contact permissions at the company's website.

(c)   **Mobile phone text messaging (SMS)**. Messages can be sent via mobile phone to a captive audience, catching them wherever they are. This form of marketing is still in its infancy, but with the proliferation of mobile phone usage it is likely to be very significant, at least in terms of numbers reached. It is also becoming increasingly sophisticated, with '3G' (third-generation)

mobile phone technology. SMS marketing is governed by the Mobile Marketing Association in the UK.

(d)   **Direct response advertising** as described above.

(e)   **Mail order**. Mail order brochures typically contain a selection of items also available in a shop or trade outlet, which can be ordered via an order form included with the brochure and delivered to the customer. Mail order extends the reach of a retail business to more (and more geographically dispersed) customers.

(f)   **Catalogue marketing** is similar to mail order, but involves a complete catalogue of the products of the firm, which typically would not have retail outlets at all. Electronic catalogues can also be downloaded on the Internet, with the option of transferring to the website for transaction processing, and on CD-ROM.

(g)   **Call centres** and **telemarketing**. A call centre is a telephone service (in-house or outsourced by the marketing organisation) responding to or making telephone calls. This is a cost-effective way of providing a professionally trained response to customer callers and enquirers, for the purposes of sales, customer service, customer care or a contact point for direct response advertising.

Definition

Telemarketing is the planned and controlled use of the telephone for sales and marketing opportunities. Unlike all other forms of direct marketing it allows for immediate two-way communication.

Global case study

A survey of recent direct mail campaigns includes the following creative ideas to overcome 'junk mail' resistance and marketing fatigue.

*   **Great Ormond St Hospital, (UK)**
    How do you make it clear that a new ultrasound scanner would mean that surgeons could avoid unnecessary surgery when diagnosing children? Send a donation request in a clear package, that asks 'If you could see inside every envelope, would you open every one?'

*   **Oroverde, Tropical Rainforest Founding (Germany)**
    How do you remind potential donors of the precarious state of the world's rainforests? Send them a paint-by-numbers kit with only one colour included: black.

*   **Genesis Energy (New Zealand)**
    How do you let customers know you're there to help them save on their energy bills? Print your energy-saving tips in fluorescent ink, so they can read them with the lights off.

*   **First Direct bank (UK)**
    How do you show customers that you're still the most thoughtful bank? Send them a single sock that they can marry to the odd one we all have in our sock drawer.

   Developing Integrated Communications Strategy

3.9 Sales promotion

Sales promotion techniques add value to a product in order to achieve a specific marketing objective.

> **Definition**
>
> The Institute of Sales Promotion (ISP) defines **sales promotion** as 'a range of tactical marketing techniques, designed within a strategic marketing framework, to add value to a product or service, in order to achieve a specific sales and marketing objective'.
>
> Sales promotion activity is typically aimed at increasing short-term sales volume, by encouraging first time, repeat or multiple purchase within a stated time frame ('offer closes on such-and-such a date'). It seeks to do this by adding value to the product or service: consumers are offered something extra – or the chance to obtain something extra – if they purchase, purchase more or purchase again.

> Assessment advice
>
> It is worth being aware of the potential for confusion between the terms '**promotion**' (used as another way of saying 'marketing communications' in general) and '**sales promotion**' (which is a specialist term reserved for the techniques involved). In an exam, especially if you are reading through questions fairly quickly, it is all too easy to answer the 'wrong' question.

3.9.1 Objectives of sales promotion

The following are examples of consumer sales promotion objectives, stated in broad terms.

- Increase **awareness and interest** amongst target audiences
- Achieve a **switch in buying behaviour** from competitor brands
- **Incentivise consumers** to make a forward purchase of your brand
- **Increase display space** allocated to your brand
- **Smooth seasonal dips** in demand for your product
- Generate a **customer database** from mail-in applications

Sales promotion objectives will link into overarching marketing and marketing communications objectives.

For example:

Marketing objective	Increase 'Brand X' market share by 2 percentage points between January and December 20XX
Marketing communications objective	To contribute to brand share gain of 2% in 20XX by increasing awareness of 'Brand X' from 50% to 70% among target consumers
Sales promotion objective	To encourage trial of 'Brand X' among target consumers by offering a guaranteed incentive to purchase

3.9.2 Consumer sales promotion techniques

Consumer promotion techniques include reduced price, coupons, gift with purchase and competitions and prizes.

The range of consumer **sales promotion techniques** can be depicted as follows.

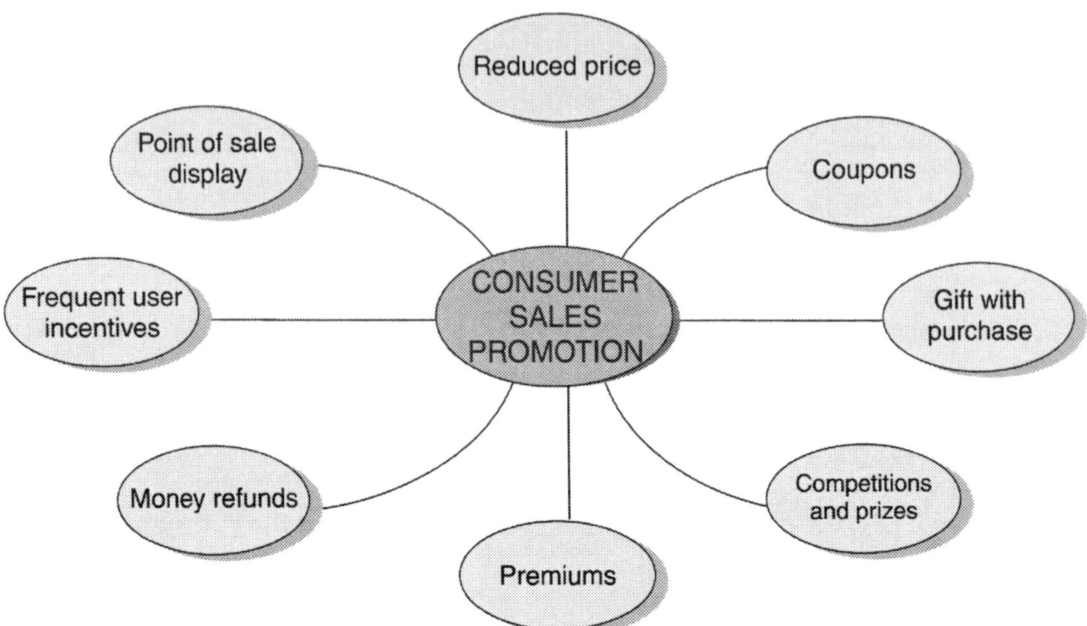

- **Price promotions**: for example discounted selling price or additional product on current purchase, or coupons (on packs or advertisements) offering discounts on next purchase

- **'Gift with purchase'** or **'premium'** promotions: the consumer receives a bonus, gift or refund on purchase or repeat purchase, or on sending in tokens or proofs of multiple purchases

- **Competitions and prizes** for example, entry in prize draws or 'lucky purchase' prizes, often used both to stimulate purchase (more chances to win) and to capture customer data

- **Frequent user (loyalty) incentives** for example, Air Miles programmes, points-for-prizes cards

Global case study

It is reckoned that two-thirds of purchases result from in-store decisions. Attractive and informative point-of-sale displays are therefore of great importance in sales promotion.

Point of sale materials include product housing or display casing (such as racks and carousels), posters and leaflet dispensers. Their purpose is to:

- Attract the attention of buyers
- Stimulate purchase in preference to rival brands
- Increase available display and promotion space for the product
- Motivate retailers to stock the product (because they add to store appeal)

3.10 Public relations

Public relations aims to enhance goodwill towards an organisation from its publics. According to Jobber (2007) it 'creates an environment in which it is easier to conduct marketing'.

Definition

The Institute of Public Relations has defined **public relations** as 'the planned and sustained effort to establish and maintain goodwill and mutual understanding between an organisation and its publics'.

This is an important discipline, because although it may not directly stimulate sales, the organisation's image is an important factor in whether it attracts and retains employees, whether consumers buy its products/services, whether the community supports or resists its presence and activities and whether the media reports positively on its operations.

An organisation can be either reactive or proactive in its management of relationships with the public.

- **Reactive public relations** is primarily concerned with the communication of what has happened and responding to factors affecting the organisation. It is primarily defensive.

- In contrast, **proactive public relations** practitioners have a much wider role and thus have a far greater influence on overall organisational strategy. The scope of the PR function is much wider, encompassing communications activities in their entirety.

3.10.1 Scope of public relations

Organisations will have to deal with more than one public, including consumers, business customers, employees, the media, financial markets and wider society.

The scope of PR is very broad. Some frequently-used techniques are as follows.

(a) **Consumer marketing support**

- Consumer and trade press releases (to secure media coverage)
- Product/service literature (including video and CD-ROM)
- Special events (celebrity store openings, product launch events etc)
- Publicity 'stunts' (attention-grabbing events)

(b) **Business-to-business communication**

- Corporate identity design (logos, liveries, house style of communications)
- Corporate and product videos
- Direct mailings of product/service literature and corporate brochures
- Trade exhibitions and conferences

(c) **Internal/employee communications**

- In-house magazines and employee newsletters (or intranet pages)
- Recruitment exhibitions/conferences
- Employee communications: briefings, consultation, works councils

(d) **Corporate, external and public affairs**

- Corporate literature
- Corporate social responsibility and community involvement programmes: liaison with pressure and interest groups
- Media relations: networking and image management through trade, local, national (and possibly international) press
- Lobbying of local/central government and influential bodies
- Crisis and issues management: minimising the negative impacts of problems and bad publicity by managing press/public relations

(e) **Financial public relations**

- Financial media relations
- Design of annual and interim financial reports
- Facility visits for analysts, brokers, fund managers
- Organising shareholder meetings and communications

3.11 Exhibitions

There are differing views on the value of exhibitions, but Jobber (2007) writes:

"... exhibitions appear to be an important part of the industrial promotion mix. One study into the relative importance of promotional media placed exhibitions as a source of information in the industrial buying process second only to personal selling, and ahead of direct mail and print advertising."

Shows, exhibitions and trade fairs offer numerous marketing opportunities.

- **Public relations** either through visitors or media coverage
- **Promoting and selling** products and services to a wide audience of pre-targeted potential customers, particularly where demonstrations or visual inspection are likely to influence buyers
- **Networking** within the industry and with clients
- **Testing the response** to new products
- **Researching competitor products** and promotions
- **Researching suppliers' products** and services and making contacts along the supply chain

Most industries are catered for by at least one annual or bi-annual exhibition in the UK, as well as internationally. The events themselves are set up by exhibition organisers who are responsible for booking and preparing the venue, registering participants and organising seminars and events, issuing catalogues of standholders and events, providing stand construction services, and organising facilities, promotions and press coverage. Exhibiting is a demanding and expensive exercise, which should only be undertaken after research and cost-benefit analysis.

Global case study

New media is having an impact on the exhibitions industry. The Internet can be used to attract visitors to shows and for post-show marketing via email.

In addition, some exhibitors are using webcast technology to broadcast their shows live on the Internet. While this is not the same as being there, such facilities are still useful for communicating information about exhibited products to those who cannot attend the show. Online visitors can view just the stands that they are interested in.

The Director General of the Association of Exhibition Organisers says 'The growth of exhibitions will go hand in hand with the growth of electronic media'. Visitors can use the Internet to book exhibition tickets, flights, accommodation and meetings.

3.12 Sponsorship

Definition

Sponsorship involves supporting an event or activity by providing money (or something else of value, such as prizes for a competition), usually in return for naming rights, advertising or acknowledgement at the event and in any publicity connected with the event. Sponsorship is often sought for sporting events, artistic events, educational initiatives and charity/community events and initiatives.

Sports sponsorship is by far the most popular sponsorship medium as it tends to offer high visibility through extensive television and press coverage.

Sponsorship is often seen as part of a company's socially responsible and community-friendly public relations profile. It has the benefit of positive associations with the sponsored cause or event. The profile gained (for example in the case of television coverage of a sporting event) can be cost-effective compared to TV advertising, for example. However, it relies heavily on awareness and association. Unless additional advertising space or 'air time' is part of the deal, not much information may be conveyed.

Marketers may sponsor local area or school groups and events – all the way up to national and international sporting and cultural events and organisations. Sponsorship has offered marketing avenues for organisations which are restricted in their advertising (such as alcohol and tobacco companies) or which wish to widen their awareness base among various target audiences.

- There is wide corporate involvement in mass-support sports such as football and cricket.

- Cultural sponsorship (of galleries, orchestras or theatrical productions) tends be taken up by financial institutions and prestige marketing organisations.

- Community event sponsorship (supporting local environment 'clean-up' days, tree planting days, charity fun-runs, books for schools programmes) is often used to associate companies with particular values (for example, environmental concern, education) or with socially responsible community involvement.

3.12.1 The purpose of sponsorship

The objective of the organisation soliciting sponsorship is most often financial support – or some other form of contribution, such as prizes for a competition, or a prestige name to be associated with the event. In return, it will need to offer potential sponsors satisfaction of *their* objectives.

The objectives of the **sponsor** may be:

- **Awareness creation** in the target audience of the sponsored event (where it coincides with the target audience of the sponsor)

- **Media coverage** generated by the sponsored event (especially if direct advertising is regulated, as for tobacco companies)

- Opportunities for **corporate hospitality** at sponsored events

- **Association** with prestigious or popular events or particular values

- Creation of a **positive image** among employees or the wider community by association with worthy causes or community events

- Securing **potential employees** (for example, by sponsoring vocational/tertiary education)

- **Cost-effective** achievement of the above (compared to, say, TV advertising)

- Creation of **entertainment opportunities** (in the sample case study, Innocent Drinks sponsors an annual music and comedy festival in London)

Sponsorship as a promotional technique also has limitations.

- Sponsorship by itself can only communicate a restricted amount of information (unless integrated with advertising and other initiatives).

- Association with a group or event may also attach negative values (such as sports-related violence and alcohol abuse).

List some examples of sporting, artistic, educational and community sponsorships that you are aware of in your country (or internationally).

- What image of the sponsoring company or brand does association with that particular event/group/cause create?

- How much promotional coverage (advertising, publicity) does the sponsor get as a result of sponsorship: how much information about the organisation or brand is conveyed?

3.13 Online forums, blogs and social networks

Web 2.0 is a term describing the trend in the use of Internet technology for purposes of creativity, information sharing and user collaboration. The recent and explosive development of web-based communities and hosted services, such as social-networking sites (Facebook, MySpace) and blogs has accelerated this trend.

Facebook, the hugely popular social networking site, was begun in a Harvard University dorm room. In October 2007 Microsoft paid $240m for a mere 1.6% stake. Facebook hopes to become an advertising magnet by substantially increasing its current audience of nearly 50 million active users.

Fail to take your customers' online habits into account and you may regret it! Firms have long sought their customers opinions on products and services, but now those opinions are showing up on the web. Any experience somebody has with a firm – usually negative – can be posted on the web for all to see. This is going to prove a massive challenge for CRM over the next several years.

People are increasingly conditioned to having their opinions and thoughts listened to, and regard companies who ignore them as arrogant and 'last century'. According to BT customer experience consultant Dr Nicole Millard, savvy customers can be a boon to CRM. Customers were calling into BT contact centres asking how to use gaming consoles with their broadband service. Agents in the contact centres didn't know the answer, but other customers had already solved the problem. So BT started a web-based forum called Hubbub, which allows customers to share solutions with other users, and provide contact centre staff with another source of information. This was a benefit both for the customers and BT staff.

4 Evaluating and measuring communications plans

4.1 Measuring effectiveness

To succeed in achieving their goals, communications must:

- Gain attention
- Communicate a message
- Improve attitude to the brand
- Reinforce the already positive attitude to the brand
- Obtain the readers'/listeners'/viewers' liking for the message and its execution

Developing Integrated Communications Strategy

Ideas-storm a list of media which might be suitably targeted (by factors in the media themselves, or in the media habits of the target audiences) for effective advertising of any of the following products/services.

(a)   A local garage offering car service, maintenance and parts
(b)   An up-market restaurant
(c)   A software package for use by accountants (based on UK law and regulation)
(d)   A new brand of washing powder
(e)   A microchip for use in engineering applications

4.1.1 How to assess the effectiveness of communications campaigns

Principle	What it means
Value linkage	Communication must represent the value of the brand.
Sense making	Communications must be meaningful and relevant.
Simplification	Strict simplification is necessary in view of the avalanche of information.
Acceleration	The message must be transmitted in a few seconds.
Visualisation	We must communicate visually first and foremost.
Humanisation	Communication must relate to the lives and dreams of real people.
Emotionalisation	Communications should activate the receiver's feelings more.
Conditioning	Effective use requires strong, unambiguous, vivid stimuli.
Refreshment	Receivers can become bored or tired of campaigns.
Branding	The brand has to be a coordinated part of communications.
Entertainment	Entertainment can be extremely effective in some cases.
Consistency	Consistency has been the mainstay of major brands.

The difficult part is measuring the effectiveness of the marketing communications process. The following are some possible techniques.

Marketing communications methods	Examples of measurement
Personal selling	Sales targets
Public relations	Editorial coverage
Direct marketing	Enquiries generated
Advertising	Brand awareness
Sales promotion	Coupons redeemed
Exhibitions	Contacts made

4.2 The role of market research

When measuring marketing effectiveness, three key questions exist:

1   How can we satisfy customer needs?
2   How do we ensure we are competitive within the market?
3   What external factors are likely to affect us?

To address these questions the organisation needs to gather information on customers, competitors and the marketing environment. This is **market research**. It is as important to use at the situation analysis or auditing stage of the plan.

The table below outlines the broad issues and then focuses on some of the detail that marketers will need to consider.

Customer information required	Information required on competitors and other organisations	Information required about the marketing environment
Who are our customers? Can our customers be segmented in any way? Are customers B2B or individual consumers or both? Do we have groups of key customers or a broad market appeal?	**Competitor activity** Who are our competitors? What are their core competences? What share of the market do they have? What additional threat do they pose?	**Macro environment** What are the PESTEL factors to impact us? Is our market growing or in decline? Are we likely to remain profitable in the current and future market conditions?
Where are customers found? Location? Frequent visitors to where? Online presence, sites visited? Where can we be available at their convenience?	**Performance benchmarking** Who should we measure ourselves against? What metrics should be used to provide a meaningful analysis?	**Micro environment** How effective is our internal market? Are we working with the right partners, suppliers etc? Internally are we organised as best as we can be to meet customer needs? Do our people understand our customers and how to best meet their needs?
How do we build a relationship with customers? Are our customers exclusively loyal? Are we part of a portfolio of brands that our customers purchase? How do our customers like to communicate with us? What is our history with our customers? How do customers perceive us?	**Partner organisations and marketing networks** What referral markets should we belong to? Are we making the most of the networks that we belong to? How can we partner with other organisations for mutual gain?	

How do customers make purchases?		
What is the decision-making process for consumers?		
What reference groups are important?		
What decision-making units are involved in the purchase?		
Do customers regard purchases as high or low involvement?		
How do we satisfy customers needs?		
Have we correctly identified what customer needs are?		
How does our offering meet their needs?		
What are satisfaction levels?		
How can satisfaction levels be improved?		
What are the behaviour patterns of customers?		
Do customers relate to our product/service individually or within a group?		
What is the influence of third parties on the behaviour of customers with regard to our product/service?		

1 Enhance your understanding of the concept of integrated plans

- Targeting the communications mix for stakeholders means identifying and using the appropriate communication media and the appropriate modes of communication

2 Use a planning framework to develop an integrated communications plan

- The integrated marketing plan should use SOSTAC® as a planning framework

- Objectives can be sales or communications based and are essential.

- Plans should be carefully researched and fully justified with detail of resource requirements, measurement devises and clear strategic and tactical suggestions.

3 Prepare and manage communications budgets

- Budgets co-ordinate activities, communicate targets and control plans

- Procedures should be set for managing the process

- A number of methods for setting budgets exist

4 Plan communications tools in a relevant manner within the integrated plan

- Media neutral planning suggests that communications should be planned according to the needs of the message and that there should not be a preference shown for one or more media from the outset.

- Promotional tools

5 Evaluate and measure the success of the integrated plan

- A range of metrics can be used to measure communications plans, many of which are based on measuring against the objectives of the campaign.

- Marketing research should be employed

1   Your answer should reflect the needs of the target audience

2 - 5 The remaining activities will depend on your own research or product chosen.

De Pelsmacker, P., Geuens, M. and Van den Bergh, J., (2006). *Marketing Communications.* 3rd edition. Harlow: Pearson.

Fill, C., (2009). *Marketing Communications: Interactivity, Communities and Content.* 5th edition. Harlow: FT Prentice Hall.

Pickton, D. and Broderick, A., (2004). *Integrated Marketing Communications.* 2nd edition. Harlow: Prentice Hall.

Smith, P.R. & Taylor, J., (2004). *Marketing Communications: an Integrated Approach.* 4th edition. London: Kogan Page.

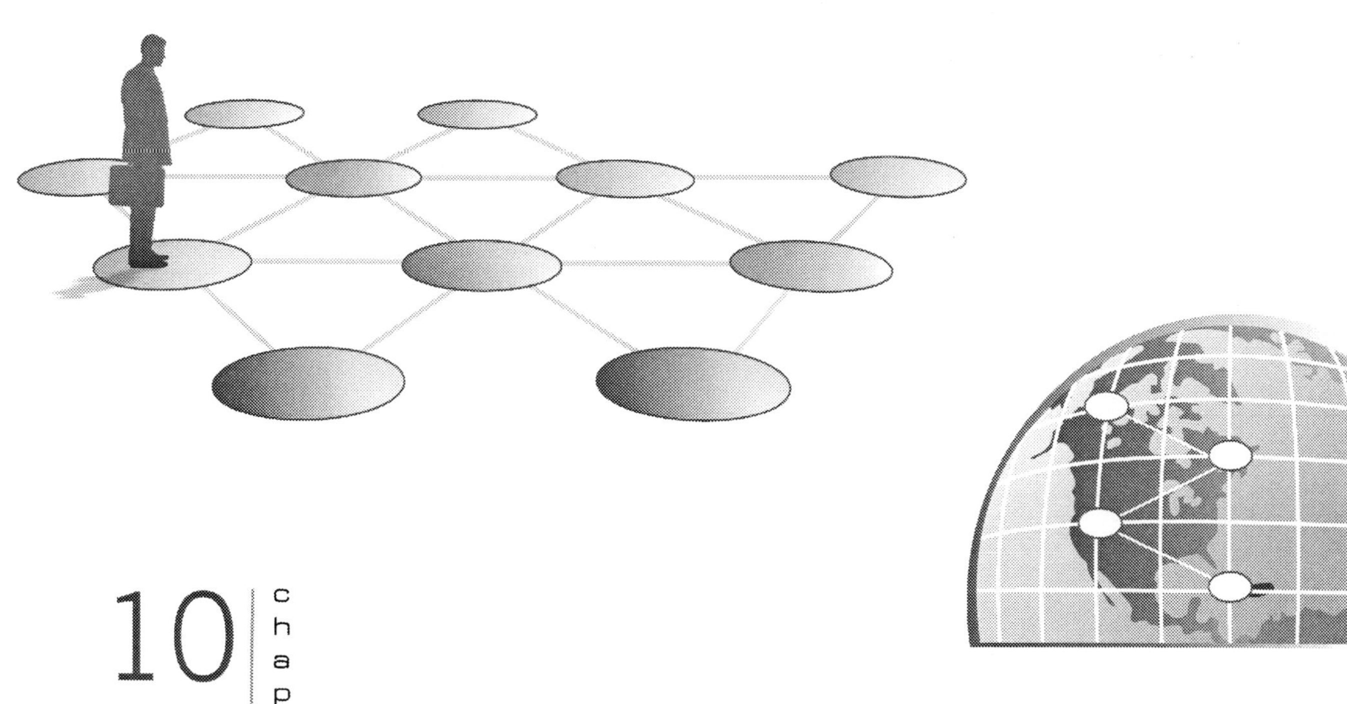

10 chapter

Managing the communications strategy

Within this chapter we finally move to look at how communications strategies can be implemented and managed. Although we have discussed the need for integration and the growing trend for media-neutral planning, there is still likely to be several agencies who are employed by the organisation. Careful management of those agencies is required to ensure that there is a consistency in the brand interpretation.

Similarly this consistent approach also needs to be taken when communicating with a large variety of global audiences. Business to business and internal communications also warrant specific attention within this final chapter to consolidate your learning from the other parts of the Study Text.

The chapter begins with a review of the implementation process, then moves to consider agency relationships. We also take the opportunity to consider holistically differences in global communications remembering our earlier discussions about global branding. We then look specifically at the business to business sector. The chapter closes by looking at internal communications. Internal communications are critical because internal views can make or break entire communications strategies.

Contents

Media, creative and production project implementation
Agency selection and management
Global communications
Business to business communications
Internal communications

In this chapter you will cover the following:

- Managing agencies and implementing communications plans
- Providing consistency with various global communications audiences
- The differences within business to business communications
- How to manage internal communications

1 Media, creative and production project implementation

Pickton and Broderick (2004) liken implementing communications projects to being a conductor in an orchestra. The different communications specialists need to work together just as musicians do in order to play a symphony. Sometimes barriers arise within organisations when it comes to the implantation of the marketing plan as shown in the table below.

Barrier	Details
Lack of horizontal communication	Communication is required across functional disciplines such as marketing, sales, merchandising, digital departments
Functional specialism	Marketing communications tends to be fragmented with specific specialists eg creative and planning staff etc
Decentralisation	Empowering employees does not automatically make them managers
Lack of corporate direction and communication	Too little corporate focus even where there is a defined mission and strategy
Lack of IMC planning and expertise	Skills to manage projects of this nature or lack of IMC knowledge
Lack of budget	One of the most frequently cited reasons is that the budget proposed is not approved
Fear of change	This is a commonly cited area of blockage
Lack of database	Databases are vital to the success of IMC

Table based on Pickton and Broderick (2004).

Clearly project management skills are required.

1.1 Developing vision and shared project understanding

It is important to spend time at the earliest stage of project planning on reaching a shared understanding of what the project is about. This applies to all projects but is particularly vital to the kind of 'soft' project that will be common in marketing. Lewis (2008) recommends a process of brainstorming, discussion and group activity in order to create a sense of ownership and an agreed vision and understanding of what the project is to achieve. Lewis also recommends the use of mind-maps as an aid to this process and suggests that the outcome should effectively be a list of desired features to be provided at the culmination of the project. These features or attributes should be prioritised into three groups.

- Musts
- Wants
- Nice to haves

When this stage is complete, it is appropriate to move on to more detailed consideration of consideration of **project scope and strategy**.

Grundy and Brown (2002) summarise the process of **project definition** as the preparation of answers to a series of questions.

- What **opportunities and threats** does the project present?
- What are its **objectives**?
- What are its potential **benefits, costs** and **risks**?
- What is its overall **implementation difficulty**?
- Who are the **key stakeholders**?

A number of techniques that aid analytical thinking may be used when addressing these questions.

1.2 Stakeholders

> **Definition**
>
> **Project stakeholders** are the individuals and organisations that are involved in, or may be affected by, project activities and outcomes.

1.3 Project stakeholder interests

It is important to understand who has an interest in a project, because part of the responsibility of the project manager is **communication** and the **management of expectations**. An initial assessment of stakeholders should be made early in the project's life, taking care not to ignore those who might not approve of the project, either as a whole, or because of some aspect such as its cost, its use of scarce talent or its side-effects. The nature and extent of each stakeholder's interest in and support for (or opposition to) the project should be established as thoroughly as possible and recorded in a stakeholder register. This will make it possible to draw on support where available and, probably more important, anticipate and deal with stakeholder-related problems. The project manager should be aware of the following matters for each stakeholder or stakeholder group.

(a)   Goals
(b)   Past attitude and behaviour
(c)   Expected future behaviour
(d)   Reaction to possible future developments

1.4 Managing project stakeholders

It is important to be thorough in assessing stakeholders: a failure to identify and consider a stakeholder group is likely to lead to problems. Here is a list of individuals and groups that might be expected to have a stake in a project.

(a)   **The project manager** – reputation and even employment are at stake.

(b)   **The project team** – they have to work on the project but may have other roles and higher personal priorities.

(c)   **The project sponsor/ budget holder** – this senior figure will demand high standards but may be reluctant to fund all the desirable resources.

(d)   **The customers** – may be external or internal and may break down into several groups, such as an internal user group and an external group who should receive improved service as a result of the enhanced capability the user group should acquire on successful completion of the project.

(e)   **Managers and staff in supporting functions** such as finance and HR can have extensive influence on the project and its chances of success.

(f)   **Line managers of seconded staff** – their own work and prospects may be disrupted.

(g)   **Other project managers, teams and sponsors** – rivalry for resources may cause conflict.

(h)   **Regulators and auditors** of all kinds, both internal and external, are likely to take an interest in a project.

Project stakeholders may be divided into two main groups.

(a)   **Process stakeholders** have an interest in how the project process is conducted. This group includes those involved in it, those who want a say in it, and those who need to evaluate and learn from it.

(b)   **Outcome stakeholders** have an interest in the outcomes, results or deliverables of the project. This group includes not only the customer groups described above, but also, typically, some regulators and compliance staff.

The project manager must manage a broad range of complex relationships if the project is to succeed. This amounts to achieving co-operation through influence rather than authority. Gray and Larson (2008) speak of creating a social network to manage the relationships with those the project depends on. The project manager should know whose co-operation, agreement or approval is required for the project to succeed and whose opposition could prevent success. It is particularly important to enlist the support of top management. This not only brings an adequate budget, it signals the project's importance to the organisation and enhances the motivation of the project team.

1.5 Managing stakeholder disputes

Unfortunately, even the most skilled and diplomatic project manager is likely to encounter disputes among project stakeholders. All project managers should be aware of the potential for disputes around their projects and be prepared to act to resolve them.

Key external stakeholders associated with the implementation of communications strategies are agencies.

2 Agency selection and management

2.1 Using an agency

2.1.1 The role of marketing agencies in marketing communications

According to Jobber (2007) an advertiser has four options when developing a campaign.

(a)   Small companies may develop advertising in co-operation with people from the media. Copy might be written in-house, for example, with artwork and final layout designed by the publication

(b)   The advertising function may be carried out in-house, but its media buying power will be low

(c)   Many advertisers opt to work with an advertising agency because of the specialist skills that are needed

(d)   A fourth alternative is to use in-house staff for some of the work and specialists for the rest. Such specialists often hold large buying power, or special creative talent in the form of 'hot shops'.

Most companies which advertise use an advertising agency. There are a number of reasons why a company might want to use **external marketing services**, rather than plan or carry out marketing activities in-house.

(a)   Professional marketing agencies specialise in communications, media and consumer trends, and stay up-to-date.

   Developing Integrated Communications Strategy

(b)   They have wide experience with a range of clients, which they can bring to bear on problems and opportunities.

(c)   They supply specialist organisational and creative talent with specialist facilities (for example, for filming, sound recording and editing).

(d)   They are objective about the client's needs and position, and may be better able to view the product from the consumer's point of view.

(e)   The cost of using external services, past a certain level of sophistication, is usually cheaper than attempting to perform the same functions in-house.

In the last few decades, marketing techniques have increased dramatically in range and complexity. Standard 'above-the-line' marketing (where media space is paid for), handled by large advertising agencies, has been augmented by 'below-the-line' (promotional) activities. Agencies specialising in sales promotion, direct marketing, e-marketing and PR have sprung up as independent operations and divisions of large advertising agencies. Many companies now use more than one agency to handle a portfolio of techniques above- and below-the-line, or to cover local and on-going work as well as national strategic campaigns.

In addition, the complexities of new media (such as e-marketing) have created technical specialisms such as web page design, data-mining, and digital media buying.

2.2 Marketing service providers

Here we will briefly outline some of the service providers with whom a marketer may have to deal.

2.2.1 Market researchers

Market researchers can be commissioned to gather and analyse information on any aspect of the business, market or environment.

2.2.2 Advertising agencies

Advertising agencies are still the most commonly-used marketing service. Agencies come in a range of sizes and specialisms to reflect an increasingly fragmented and diverse range of media.

(a)   'Media only' service (where the agency selects, schedules and books advertising space at a discount, which is then passed on to the client)

(b)   'Full service' (where the agency creates the whole campaign from concept through production to press/air)

Many larger, integrated agencies now have divisions specialising in various areas:

- Sales promotion
- Direct marketing
- Public relations
- E-marketing (Internet, SMS, podcasts etc)
- Design

2.2.3 Promotion agencies

Promotional activities – collaborations between brands, competitions and incentives – have been a huge growth area in marketing. Some agencies specialise in these areas:

- Sourcing promotional incentive products and merchandise
- Devising links and negotiating deals between brands
- Organising competitions
- Designing and producing promotional packaging and information material

PR consultancies handle several areas.

(a)   **Media or press relations** – keeping the media informed, in order to manage the company's image and secure favourable coverage

(b)   **Corporate relations** – promoting a corporate image to the public, market and business world

(c)   **Marketing support** – promoting specific products or services via publicity, events and press coverage

(d)   **Government relations** – lobbying on behalf of the company's interests

(e)   **Community relations** – targeted at the general public or local residents, via communication and social programmes, community involvement, sponsorship

(f)   **Financial relations** – communicating with shareholders, financial media, the stock exchanges

(g)   **Employee relations** – communicating with staff

PR agencies are less easy to manage than advertising agencies, because their activity and output is often less tangible. The major cost is their time plus expenses. In order to make cost effective use of their services a marketer must consider the following:

- Set clear objectives for the plan or project
- Set a project fee where possible
- Brief comprehensively
- Monitor activity
- Monitor expenditure

2.2.4 Direct marketing services

Direct marketing agencies offer a range of services from database and mailing list development, analysis and segmentation to telemarketing, direct mail package design, copywriting, mailing and fulfilment – and their electronic equivalents such as email and SMS.

2.2.5 Brand identity

There are specialist agencies which perform specific brand-related services:

- Designing and producing product packaging and display
- Orchestrating brand updates
- Researching, devising, testing and registering new brand names

2.2.6 Sponsorship

Agencies undertake negotiations for:

- Sponsorship agreements
- Product endorsements by celebrities
- Product placement in films or TV programmes

2.2.7 Creative services

Agency, freelance and corporate help is also available with:

- Design and layout, artwork and illustration
- Concept development and copy writing
- Print buying and print production
- Photography
- Film, video and audio production
- Web page design

In addition to all of the above, the marketing department may deal with many providers of services related to promotional events and activities:

- Event organisers, venue managers, hotel and restaurant staff
- Exhibition organisers, stand constructors and removalists
- Postal services
- Model/talent agencies, costumer hirers etc

Activity 1

The table below shows two small ads. What sort of companies/products might use these services?

Market Research Limited		CHILDREN AND YOUTH
Trade Research Specialists		BRITAIN AND CONTINENT
• Retail audits studies • Product availability surveys • Mystery shopping • Customer interviews • Consumer research	• Tailor made • Overnight pricing checks • Shop testing • Trade interviews	**GB** *Core Sample:* 1,200 7 - 19 year olds with bi-monthly extensions to include ages 3 to 6 and 20-24. Any age range within these limits. *Field dates:* every 2 - 4 weeks. *Rates:* from £290 per question according to age range covered. **Continent:** Five country child and youth surveys

2.3 How agencies are structured

2.3.1 Who's who in the agency?

The simplest form of **agency structure** is as follows.

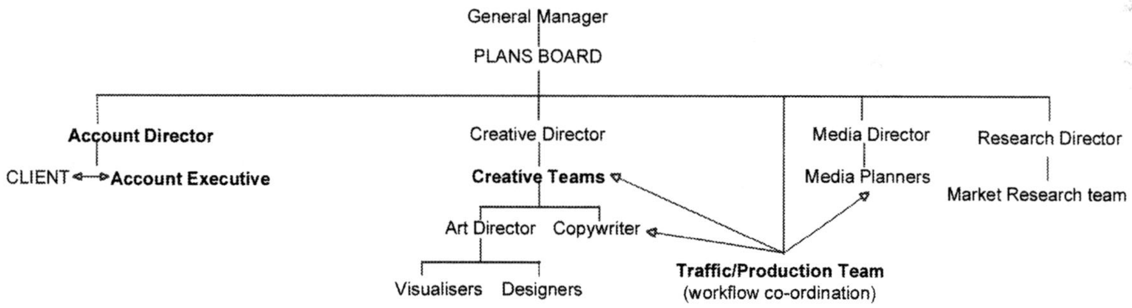

- The **account executives** service the clients on a day to day basis

- The **account director** monitors work on given accounts

- The **creative team** 'produces' the ad

- The **traffic/production team** plans and schedules agency workflow, monitors progress against deadlines and 'chases' work

The account director/manager is vitally important because a full service agency will provide the services of not fewer than 20 departments, and these need to be co-ordinated around a client focus. His role is to liaise with client staff and to brief, supervise and co-ordinate the appropriate agency staff at appropriate times. The departments commonly found in a **full service agency** are as follows.

• Creative	• TV and radio production	• Marketing
• Typography	• Press production	• Media planning
• Presentation	• Press buying	• Market intelligence
• Studio	• Print buying	• Personnel
• Account management	• Art buying	• Finance
• Economic forecasting	• TV and radio buying	• Administration
• Research	• Sales promotion	• Management

2.3.2 Internal structure of an agency

The internal structure of an agency can best be explained by considering how the agency handles a piece of client work.

(a) **The client problem**

A client and agency who have been together for some time will have built up a good working relationship. The agency will understand the client's business, and the motivations and decision-making processes of end consumers. A new product or service, new situation or changing market conditions may provide the starting point for a new role to be performed by advertising. The client needs to brief the account executive on the task in hand.

(b) **The internal briefing**

The account executive will brief the members of the account team who work on the client's business. These members are:

(i) An **account planner**, responsible for using market research to develop advertising strategy for the clients.

(ii) A **creative team** or duo of art director and copywriter, responsible for conceiving a creative idea which meets the advertising brief and working that idea up into visual form and written copy.

(iii) A **media planner/buyer**, responsible for recommending an appropriate media strategy and ensuring media is bought cost effectively.

(c) **The client presentation**

The account executive will present back to the client. He will show examples of how the final advertising execution will look, using rough visuals or storyboards and will explain the rationale for the ideas presented. Depending on the client's reaction to the team's interpretation of the brief, the team will either be asked to go ahead in developing the work, or will be asked to rework their ideas.

The go-ahead stage may include a decision to test the advertising in research prior to full production of the advert.

(d) **Production of advert(s)**

Some simple advertising executions will be carried out almost entirely in-house by the advertising agency. For more complex executions, the agency will buy in specialist functions on behalf of the client.

While production is ongoing, the media department will be involved in the actual commitment of the media budget. Dates, times and positions will be agreed with media owners.

During this stage, there will be continuous liaison between the account executive and the client. The client is likely to attend some of the key production stages such as filming or photography of the commercial.

If time allows, further research may take place to identify the need for any additional changes.

(e)   **The campaign appears**

The time span between the briefing of the account executive by the client and the campaign appearing can be as little as six or eight weeks for a simple photographic newspaper execution, to 20 weeks plus for an animated TV commercial. Once the campaign has appeared, it is important that it is properly evaluated against the objectives initially set.

Members of the account team, who work directly on the client's business in creating a campaign, are those the client is most likely to come into contact with, although the main point of contact will of course be the account executive. There are other behind the scenes staff with whom the account executive must liaise but who will not have direct client contact.

- **Accounts department**, responsible for billing the client and paying agency invoices
- **Vouchers department** checks that press adverts appear
- **Traffic department** ensures that jobs are taken through their different stages on time.
- A large agency may additionally have an **information** or **library service**, and **legal department**.

An IPA (*Institute for Practitioners in Advertising*) booklet 'Getting the most out of the client-agency partnership' suggests the following checklist of questions.

1   Are the client's **corporate objectives**, and the **marketing objectives** that derive from them, entirely clear to both parties? Have the marketing and communications tasks been quantified, together with their financial and profit implications?

2   Are the marketing and communications planners fully informed about **relevant factors in the commercial environment**, particularly:

- Actual and potential customers
- Product advantages and limitations
- Competitors
- Current or predictable marketing problems
- Product and marketing development plans?

3   Has sufficient **lead-time** been allowed for planning purposes?

4   Has the agency's **creative policy** been fully thought through and discussed in relation to the defined communications tasks?

5   Do the various creative manifestations (whether advertisements in paid media, explanatory literature, direct mail shots, audio-visual and other sales aids, exhibition stands or press publicity) **express the policy** effectively?

6   Have the **media plan and budget** been fully thought through and discussed? Are both agency and client convinced that they are as cost-effective as available information will permit?

7   If **results** in one area have not come up to expectations, has there been a serious effort to find out *why*, instead of shrugging it off or explaining it away?

8   Is there sufficient readiness not just to change for change's sake, but progressively to **improve performance** within an agreed long-term policy?

9   Is the level of **financial and administrative control** satisfactory on both sides?

10   Is the level of **client-agency communications** – in both directions – as good as it should be?

11   Is there **mutual understanding and respect** between the individuals working on the account, on the client and on the agency side?

2.4 Agency selection

2.4.1 The selection procedure

The stages of the selection procedure are as follows.

- Define the campaign requirements
- Develop a list of suppliers
- Assess their credentials
- Issue the brief to a short list of candidates

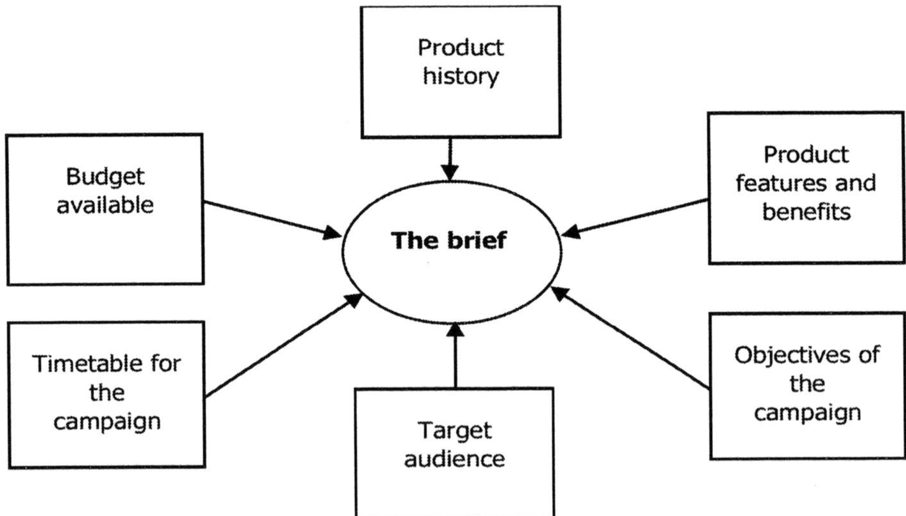

- Presentation
- Analysis
- Selection of the chosen agency
- Agree contract details

Activity 2

Explain the significance of the elements of the creative brief.

2.4.2 Criteria for assessment

It is essential that each agency is judged by the same criteria. The criteria should be established by the client in advance, and they must be understood and agreed by all who will be part of the selection team.

Criteria for assessing an agency	
Item	Comment
Present work	• Is it exciting/interesting? • Is it effective? What proof is there? • It is allied to our marketplace?
Present clients	• Are there any clashes that will worry us or the agency or the other client(s)? • Do previous clients come back for more or are agency/client relationships short lived?
Chemistry	• Will we feel able to trust the agency? • Will the trust be reciprocated?

 Developing Integrated Communications Strategy

Item	Comment
Staffing	• Will the people who worked on the pitch work on the account? • How stable will our account team be? • What depth of experience can the account team offer?
Evidence	• Have we seen evidence of creativity, of production, of media planning, of management, of the particular skills we need?
The pitch	• Is it a package or tailored specially? • Does it arouse us to buy?
The agency	• Is its size, age and structure suitable? • Does it have to have international capacity? • Is it a full service agency or a specialist? • What is its workload? Does it have time for us? • Does it possess the necessary business skills? For example, has it negotiated a fee for specialised work necessary to meet the requirements of our invitation to pitch for the account?)
Specialised knowledge	• Has it demonstrated realistic and satisfactory understanding of our organisation and market?
Price	• Does the asking fee represent value for money when the above points are taken into account? • Does the agency know its value to us?
Judgement	• Is this the agency for us?

2.4.3 Final selection

The final selection will involve a judgement about how well the client believes the agency has responded to the brief in terms of the strategic thinking involved, the creative work (if presented) and the agency's all round understanding of the client's industry. Courtesy dictates that the client should inform both the successful agency and the unsuccessful ones in a prompt manner.

2.4.4 Criticism of the formal selection process

The formalised process is now being questioned, especially because with the arrival of new media and Internet technology, campaigns are having to be resolved in days rather than weeks. The pitching process can lead to tensions, especially on the part of unsuccessful agencies who may have invested a great amount of time and effort for no reward. Perhaps most significantly for them, they will have shared their ideas, over which they will have no subsequent control.

The pitching process also gives little insight into subsequent working relationships.

2.5 Fees and fee structures

2.5.1 Payment of creative agencies

Historically, agencies have earned their money through commission on media space purchased for their clients. The practice arises from the time when the advertising agent was a media broker who also provided other services. This method of payment also highlights the agency's legal standing. The agency is liable for bills to the media if the client defaults on payment. Nearly 50% of agency income still comes from commission. Retainers are also sometimes paid to agencies by a client. A retainer will ensure that the agency is available to comment on behalf of the client and there may be a base rate of work agreed in

return for the agreed monthly fee. PR agencies are most likely to be employed on a retained basis. Clients increasingly try to avoid retainer based arrangements.

2.5.2 Agency commission

15% used to be the standard rate of agency commission. This is now no longer the case. Some large clients have argued that they should pay the agency a discounted rate of commission, because of the volume of media throughput that the agency handles. Other clients, themselves under pressure to make advertising money work harder, have argued for commission rates of 10% or 13%.

2.5.3 Fee payment

Some advertisers and their agencies prefer to work on a project by project fee system. This ensures the agency earns money, whether or not the work is media based. About 40% of agency income is earned in this way.

Global case study

David Ogilvy is on record as saying 'I pioneered the fee system but I no longer care how I get paid, providing I make a reasonable profit. With a fee system the advertiser pays only for the services he wants, no more and no less. Every fee account pays its own way. Large profitable accounts do not subsidise small, unprofitable ones. Cuts in client's budgets do not oblige you to cut staff. When you advise a client to increase his advertising he does not suspect your motive.'

It is essential for the agency to have an efficient system to capture accurately and promptly all data relating to the allocation of staff time, and the utilisation of other resources. It is reasonable for a client to expect to be shown the control method in use, but not to have access to the agency's detailed profitability. A typical fee calculation is shown below.

Advertising agency: typical fee calculation

	Actual hours	Amount £
*Direct time cost**		
Account management	1,000	40,000
Creative	1,500	60,000
Media	300	12,000
Production	400	16,000
Research and planning	750	30,000
	3,950	158,000
Overheads at 110% (including secretarial, managerial, accounting and administration payroll; also establishment and general costs)	-	173,800
Direct expenses (including directly attributable travel and presentation costs)	-	14,000
Total hours/cost	3,950	345,800
Gross profit at 25%	-	86,450
Total (£109.50 per hour)	3,950	432,250

* Includes direct and indirect payroll costs of all staff who allocate their time.

The sums charged per hour will obviously vary from agency to agency and over time. The hours needed, however, should remain constant. It will be seen that a new product development can occupy many hours. Even more hours will be used within the client, but they are unlikely to be controlled as rigorously. There is therefore considerable unquantified investment in any new creative development. The only party

able to account for its involvement accurately is the agency. (Unfortunately for the agency, its efficiency in this respect means that is likely to be the first target in any cost-cutting exercise!)

2.5.4 Payment by results

Payment by results schemes have been used mainly in the USA, although they are becoming more common in the UK. With performance-related payment, the agency is judged on the effect its advertising has on client company sales. Different rates of commission then come into force, depending on performance to target, over-achievement or under-achievement. The major drawback to this method of remuneration is that it pre-supposes a direct correlation between advertising effort and sales.

2.6 Managing agency relationships

2.6.1 Managing agencies locally and internationally

The specialist agencies providing support services (advertising, public relations, sales promotion, research) usually all work to much the same principles. There will be an account director directly responsible for the client's account and working directly with the client brand manager. Thus all contact between client and agency is via the conduit established by these two individuals.

Client's responsibilities

(a)   The advertising co-ordinator (or marketing department head) briefs the agency.

(b)   The marketing department head approves the agency's assessment of the advertising objective, strategy, schedule and cost estimates.

(c)   The advertising co-ordinator provides the agency with all available materials and information to help the creative team to produce effective ads.

(d)   The advertising co-ordinator is the first one to see work that comes back from the agency at each stage of production (copy, story boards, sketches), and will probably check, correct and recommend any necessary changes before the ad is finally approved by the marketing department head.

(e)   The ad is returned to the agency for typesetting, final art, filming and sound production.

(f)   The advertising co-ordinator carries out final checks, and gets final approval before giving the agency clearance to release the ad to the media.

Establishing a relationship with a creative agency takes time. The end result has to be an on-going relationship based upon mutual understanding and respect. Therefore both the selection process and the day-to-day relationships have to receive very careful attention.

A creative agency should be a full member of the client's marketing team. As such, members of the agency have to be trusted with market information and, to some extent, with profit information as well. If they are not, then information will be withheld and the relationship will suffer. The creative work will suffer too, and sales and profitability will fall. If an agency understands the costing of a package it is less likely to spend time devising a promotion that the client will not be able to afford, and so keeping working data from an agency is counter-productive, morale sapping and expensive.

The only reason an agency is engaged is because it can do a job or provide a service better (and more quickly) than the client can do it for himself. It follows that the client should not meddle with their work. If the client does not like it, of course, he must say so, but the test must be 'will it work with the target audience?'.

1   Management is by the client.

2   The agency team must work closely within the client's marketing department.

3   Briefings must be specific and unambiguous.

4   All research data and management control information available to the brand manager must be provided to the agency.

5   The agency should sit in on the client's strategic planning meetings as an equal member.

6   Time and cost requirements must be reasonable and must be accepted by the agency.

7   The client should not meddle in the creative process and the agency should not interfere with the production of the package, although both may make an input, as appropriate, as part of the strategic planning process.

8   Full credit must be given when the agency is successful, and shared responsibility must be accepted as appropriate.

9   Fees and commissions should be agreed in advance and accounts should be paid promptly.

10 Copy should never be changed once it has been approved. It should be fully checked before it is signed off. Changes after that stage are not only expensive but, more importantly, they are damaging to the brand manager's personal credibility with the agency.

Given that the agency is accepted as part of the marketing communications team it should take full responsibility for the production of cost-effective work that meets the brief and hits time targets.

International agencies

Over the last forty years, as companies have expanded their operations internationally, so too have advertising agencies. Many of the large agencies have developed internationally, either by setting up branch offices in foreign countries or by merging with or acquiring local agencies.

Despite the increasing presence and power of international advertising and media networks, **independent local agencies** exist in the markets of most countries. Some companies prefer to retain country by country agency arrangements, believing local agencies to be **creatively closer** to their own markets.

Some agencies expanding abroad prefer to establish international networks or alliances where local offices are not wholly controlled. The argument is that local partners with a stake in the agency will be motivated to produce superior work.

Media independents have mirrored the pattern of agency development and many belong to international media planning and buying groups.

The trend among clients is towards the centralisation of advertising. Many large companies believe international brands are best served by an agency operating internationally.

Selecting an international advertising agency will follow a series of well defined stages. Locating suitable agency candidates is the first step in the process.

(a)   **Initial search.** Prospective clients will probably be aware of the large multinational agencies based within their own country.

(b)   A **shortlist** of agencies will then be drawn up, usually on the basis of their current work and past track record.

(c)   The **client will then visit** the local offices of those agencies for a series of credentials presentations. These initial visits will help to form an opinion about which candidates should be requested to formally pitch for the client's business. All agencies involved in the pitch should be given the same written brief to follow.

(d)   The **agency's response to the brief** will usually involve a formal presentation backed by a written proposal document with several important features.

- The agency's interpretation of the client's advertising problem
- The creative and media strategy which will ensure objectives are met
- Control mechanisms to be used
- Timing schedules
- Allocation of responsibilities
- Costings
- Terms and conditions of business

(e)   The **final selection decision** will have to take into account many client side factors such as the client's organisational structure and management style, the number of brands to be advertised and the degree to which brands penetrate different markets.

Other criteria for selection would include the following.

- The types of advertising and other communications services offered
- Level of **expertise** in the client's field of work
- The agency's **international creative track record**
- The **balance** within the agency of campaigns handled for local clients and those handled for international ones
- Whether the agency has **strong local offices** in the client's home and other key markets
- The extent to which the agency's **culture and management** style fits with that of the prospective client
- The **potential conflict** with existing business handled within the agency network
- The **control and co-ordination** procedures in place

Your company is about to launch a group of new consumer toiletries products in the Middle East. What sorts of assurance would you want from an advertising agency pitching for this account?

The advantages of using an international agency

- Less duplication and dilution of effort on the part of agency and client
- Centralised control of all advertising effort
- Speedy response across markets
- Pooling of talent and ideas from the entire agency network
- Specialised resources available
- Standardised working methods by the agency
- Reduced costs due to economies of scale

2.7 Management and control issues

External factors, such as market diversity, segmentation and competition affect the type of co-ordination chosen. Internal factors, unrelated to the market, can also dictate the management and co-ordination of international promotions.

(a)   **Organisation structure**

(i)   **Local autonomy**. Each subsidiary of an international agency may act as a separate profit centre, attracting its own clients in the home market. Upon appointment, the subsidiary

which has brought in the client takes the role of lead agency office, with overall supervision of the client's account.

(ii) **Central control**. Alternatively, an agency may exert strong central control on regional offices from its headquarters base.

(b) **Organisation culture**. Managers may have different assumptions as to how advertising ought to be done, and it might be a basic hidden assumption that decisions are taken at the top or, on the other hand, by giving local managers responsibility.

(c) The need for **coordinated marketing communications**: firms with worldwide exposure may need central control of marketing communications to ensure they are, in fact, coordinated.

2.7.1 Current client/agency issues

As markets expand, clients are likely to forsake traditional, vertical organisational structures where brands are managed on a country by country basis, in favour of a horizontal structure which cuts across country divides. This implies brand and product management at a centralised level and may result in a preference for centralised agencies. Consequences might be as follows.

(a) **Clients** are likely to become more demanding of their agencies, as clients strive to ensure that their advertising is accountable and effective. In America, the trend is already towards payment by results. Clients are also likely to demand a larger base of expertise in terms of communications and research services provided.

(b) **Agencies** will continue to expand internationally to meet the needs of their clients. This may lead to a concentration in advertising agency ownership as the large agencies seek to expand still further by way of acquisition and merger.

(c) Agency expansion will also mean that an increasing number of local agency offices will be established in new markets (eg Russia, Eastern Europe, China). Agencies may need to take a **long term perspective** on emerging markets. Initial resource requirements will be high.

Media buying and selling power continues to concentrate. On the one hand, large international media independents hold consolidated buying power and have the ability to level volume or other discounts. On the other hand there is increasing concentration in global media ownership.

Assessment advice

Rather than thinking about advertising agencies solely from the 'outside', imagine yourself as an employee at a marketing agency that is pitching for a new contract with a global company.

Think about how the prospective client would judge your agency, how they would want to measure your performance, and how the relationship might be managed. This will help you to ensure you integrate all perspectives into your assignment.

3 Global communications

3.1 Setting the scene

Due to the amount of content we need to cover throughout this module, within the Study Text we are restricted to one chapter on the topic of global communications, and so we are really only scratching the surface of the complexity behind international marketing communications. This section begins by addressing the key question: to globalise or not to globalise? Then we go on to discuss various aspects of national differences.

3.2 Cross border marketing communications

3.2.1 A global village?

As far back as 1983, Levitt argued that consumers the world over were converging in tastes and that the globalisation of markets was at hand. Levitt saw the emerging global village as presenting a huge opportunity for multinational companies to standardise products and attract large numbers of consumers through low costs brought about by economies of scale.

Factors contributing towards the globalisation of markets include the following.

(a)   More **sophisticated consumers** who holiday outside their home country and are willing to experiment with non domestic products and services.

(b)   The trend towards elimination of political, trade and travel barriers worldwide.

(c)   The **internationalisation of broadcast and print media** (for example, cross border satellite transmission means consumers receive common programming).

(d)   The **saturation of domestic markets**, leading companies to search for growth for their goods and services in new markets.

3.2.2 Vive la difference?

Levitt's globalisation argument has been criticised for adopting a production rather than a marketing orientation. **Standardisation may suit companies, but may not be what consumers are seeking**. An alternative view is that **consumer markets are fragmenting rather than converging**. Consumers are seeking to express themselves as individuals and do not want to be treated as part of a homogeneous mass. Products and services should therefore be **adapted** to individual country markets.

These two conflicting opinions provide a starting point for any consideration of international marketing management.

(a)   It is unlikely that the majority of companies competing outside their own home market will be able to standardise their marketing mix completely.

(b)   Nor is it likely that they will choose to adapt their marketing mix totally for each country in which they operate.

(c)   The nature of the product or service, consumer buyer behaviour and competitive market environment will all dictate the appropriate strategy to adopt.

Global case study

Consumers in India are acutely price sensitive. They will think nothing of spending an entire morning scouting around to save five rupees. As a result, India has the largest 'used goods' market in the world. Most washing machines in the Punjab are used to churn butter, and the average washing machine (conventionally deployed) is over 19 years old.

"What many foreign investors don't understand is that the Indian consumer is not choosing between one soft drink and another; he's choosing between a soft drink and a packet of biscuits or a disposable razor" says Suhel Seth of Equus Red Cell, an advertising company.

What this means for foreign investors is that they must price cheaply, and therefore source almost everything locally, to keep costs down.

There are other problems. Standard refrigeration becomes pretty useless when acute power shortages occur. Most consumable goods perish pretty quickly in the climate. And the country's fragmented regional culture means advertisers have to focus on common ground (such as music, Bollywood and cricket).

Is it worth the effort? Investors say that overcoming such obstacles has equipped them for success in any market in the world.

3.3 Cultural considerations

> **Definition**
>
> **Culture** is a term used to describe the set of values, beliefs, norms and artefacts held in common by a social group.

Companies operating outside their home markets have to be aware of the implications of **cultural differences** for all aspects of the marketing mix. Marketing communicators need to be particularly sensitive to culture if messages are to work in global markets.

We absorb the culture of our home society as we grow up. Our family, our religious institutions, and our education system all play a part in passing culture from one generation to the next. Some behaviours and customs learned early in life are likely to remain resistant to the best marketing or promotional efforts. For instance, attitudes and behaviour with respect to particular foodstuffs can be culturally ingrained.

Culture evolves. Although core cultural precepts are passed from one generation to the next, the values of society do change. Before the second world war, a 'marriage bar' existed in white collar occupations such as clerical work and teaching. When a single working woman married, she was expected to give up her job to look after her husband. This social norm seems inconceivable to us today.

Various dimensions of culture are relevant to the international marketing communicator

- Verbal and non verbal communications
- Aesthetics
- Dress and appearance
- Family roles and relationships
- Beliefs and values
- Learning
- Work habits

3.3.1 Verbal and non verbal communication

Common-sense dictates that care must be taken when **translating copy** from one language to another. A catchy phrase in the home market may not work so well elsewhere. A literal translation of an advertising slogan, 'as American as apple pie' would be meaningless in many countries.

Usually, agencies will aim to interpret and adapt in order to capture the spirit of a communication message, rather than relying on a mechanical word for word translation.

Most agencies will buy in the services of a translation house when required. This will usually ensure that translation is carried out by someone who speaks the vernacular language of the country in question, an important point given that languages are in constant evolution. If an agency subsidiary or local company office exists in the overseas market, it can also be helpful to incorporate their advice. Back translation, where the copy is translated back into English, can be a final precaution.

It may be difficult to decide exactly which language to choose for translation purposes. The official language of China is Mandarin, but a large proportion of Chinese living in the south of the country have Cantonese as their native tongue. In Canada, although the majority of the population speak English, packaging must include a French translation.

An additional consideration is that of the **space required for foreign language text**.

Developing Integrated Communications Strategy

You should remember from earlier in this text that **brand and product names** must be assessed for their suitability in international markets. Mars changed the name of their Marathon chocolate bar to Snickers in 1990 to facilitate their global communications strategy. A different approach is demonstrated by companies who choose to go with different names in different markets. For instance Ford's Mondeo name works well in European markets, where the 'world' association comes across powerfully. In the US market however, the car is called the Contour.

Sometimes the phonetic sound of names can cause a problem. The French soft drink Pschitt would have obvious disadvantages in the English-speaking world.

People communicate both by using language and by their use of **non verbal signals**. A shake of the head to British people, means no and a nod yes. This convention is reversed in some cultures.

Spatial zones can communicate. In Western societies, we give work colleagues or acquaintances a large zone of personal space. Only family and close friends will enter an individual's near zone. If a colleague comes too close, our reaction is to back off, as our personal space is being invaded. This is not so in some Far Eastern countries, where space is at a premium and people have far less privacy. There, it is acceptable to stand far closer to others.

3.3.2 Aesthetics

Attitudes towards different design and colour aesthetics vary around the world. A fragrance or toiletries carton decorated with chrysanthemum flower graphics would not be a success in France, where the flowers are traditionally associated with funerals. Similarly, the colour white, which carries connotations of freshness, purity and fragility in the UK is the colour of mourning in China.

Different cultures will have grown up with their own rules about visual representation. The idea of showing a figure partially out of frame is well known in Europe. In much of Africa by contrast, figures that go over the edge of their frame transgress the cultural rule about how pictures should look.

3.3.3 Dress and appearance

Dress, be it formal or social, is very much constrained by culture. Advertisers need to be aware of cultural dress codes when deciding if an execution prepared in one market will be acceptable in another. In Europe, for instance, adverts for shower and bath products will often depict a semi-nude model. Such executions would be out of the question in the Middle East where women can only appear in advertisements if carefully attired.

3.3.4 Family roles and relationships

Family has always been the dominant agent for transmitting culture. However, it may be that family influences are on the decline, at least in Westernised societies. It is now the norm to talk about the nuclear family, geographically isolated from grandparents and other relatives. Single parent families are also on the increase. This contrasts with many third world countries, where the extended family lives in a tight knit community.

Family roles can differ greatly from country to country. Recent television adverts in the UK have shown fathers shopping in the supermarket with their children (Bisto); fathers washing children's clothes while their wives are out (washing powder commercial); and men cooking happily in the kitchen (Sainsbury's). In countries with more rigid gender codes, these executions might be either absurd or offensive.

3.3.5 Beliefs and values

Beliefs and values evolve from religious teachings, family structures and the pattern and nature of economic development in that society. In our culture, material well being is important, people tend to be defined (and define themselves) by their work roles, individualism and equal opportunities are held dear and time is a precious commodity. Different cultures may put a different emphasis on these values. Any communications imagery designed to be used internationally should therefore be scanned for **underlying values or beliefs that are culture specific.**

3.3.6 Learning

The level of education within a culture is an important factor for the international marketing communicator. Low **literacy levels** will mean that verbal methods of communication take precedence. Press advertising and direct marketing may have to be ruled out. Packaging may need to be simplified, and point of sale and sales promotion techniques handled with care.

In some societies, education emphasises the value of rote learning. In others, critical thinking is encouraged. This may affect the way in which information is transmitted within a market.

3.3.7 Work habits

Not all societies conform to the Monday to Friday, 9.00am till 5.00pm work routine. In Hong Kong, the working week extends until Saturday lunchtime. In some parts of South Europe, it is usual to work from very early in the morning until early afternoon. Workers then go home for a siesta.

3.3.8 Advertising culture

It has been suggested that in addition to being sensitive to a country's culture in general, **marketers should be aware of the level of a country's advertising literacy**. The idea is that advertising in any national culture develops in a predictable way, and according to its position on this development curve, it can be described as having either high or low advertising literacy. Thus, imported advertising can be inappropriate because the advertising of country A is at a different stage of development from that of country B.

Five levels of advertising development

Level	Comment
Least sophisticated	The emphasis is on the **manufacturer's description** of the product. Messages are factual and rational with much repetition. Product or pack shots take prominence.
Unsophisticated	Consumer choice is acknowledged so emphasis switches to the **product's superiority** over the competition (eg products that wash whiter, feel softer).
Mid point	**Consumer benefits are emphasised**, rather than product attributes. Executional devices may include the use of celebrity endorsements or role models may give demonstrations, for example a dentist endorsing toothpaste products.
More sophisticated	Brands and their attributes are well known, so need only **passing references** (perhaps by way of a brief pack shot or logo). The message is communicated by way of **lifestyle narrative** (eg Gold Blend couple; Bisto family).
Most sophisticated level	The **focus is on the advertising** itself. The brand is referred to only obliquely, perhaps at a symbolic level (eg Silk Cut; Benson & Hedges). Consumers are believed to have a **mature understanding of advertising,** and are able to think laterally in order to decode messages.

Choose a country which is quite dissimilar to your own. You could perhaps choose somewhere that you have visited on holiday, or that a friend, relative or colleague has visited and told you about. Using the cultural consideration headings we have just covered, compile a list of similarities and differences between the culture of the foreign country and that of your own.

3.4 Media considerations

Assessment advice

Media is a complex, highly specialised area within marketing communications. International media is even more of a minefield. For the Global Marketer Programme we appreciate that students come from many international markets and so you will all be likely to have a slightly different perspective.

What can be expected is a grasp of the factors influencing media choice in overseas markets, and some understanding of the practical problems which may be experienced when planning the use of media irrespective of your own country experience.

The UK is a media rich country which offers a great variety of choice to the advertiser. Press is a well segmented sector with over 6,000 consumer and trade and technical magazines, as well as the national and regional newspaper titles. Commercial television now includes terrestrial and satellite channels. Radio as a sector is strengthening, with both local and national stations. Cinema and outdoor are mature mediums.

This diversity is not always mirrored in overseas markets. Media availability can vary greatly from country to country.

Trying to gain a perspective on the international media scene is a frustrating task. Different sources give conflicting information because of different definitions and different methods of data collection that have been used. Any facts that are collected tend to become dated very quickly because of the dynamic nature of the media sector. Whatever information is gained from secondary sources will always need to be double checked for accuracy in the home market itself.

3.4.1 Press

The first point to check when contemplating press as a media option, is the **literacy levels prevailing within the country**. The majority of European nations have literacy rates of 98% or 99%.

In the UK, there are a large number of **national newspaper** alternatives (24 at the last count). Although some countries are similarly well served with national papers from which to choose (eg Germany, Finland), others have more of a tradition of reading strong daily regional papers.

The UK has one of the world's largest **trade and technical press sectors**, with business-to-business advertisers being particularly fortunate in the choice available to them. A strong trade press sector however, tends to go hand in hand with a strong economic industrial base. Less developed nations may have few, if any, trade media vehicles.

It is wise to preview any candidate title before going ahead with a decision to include it on the media schedule. This will help in gaining an understanding of the editorial stance of the publication and give an impression about the calibre of the competitive advertising. It also allows the intending advertiser a chance to gauge the quality of media reproduction.

There is some debate about whether truly **international press options** exist. Serious business journals such as *Time*, *Newsweek*, and the *Economist* are widely available, as are women's titles with different national editions (eg *Marie Claire*, *Cosmopolitan*, *Elle*). General interest magazines such as *Reader's Digest* and in-flight magazines reach many markets. However, such publications are likely to be read by only a niche business and professional or lifestyle group. To reach the less wealthy, less well travelled wider population, other local media will have to be accessed.

In Germany, where more advertising money is spent in magazines than in any country other than in the USA, the main classifications are supplements, 'ad mags', consumer magazines, customer magazines, and trade and technical journals. Supplements are loose inserts in newspapers or periodicals with a magazine format and with many pages printed in four-colour. The principal types are programme, quality, special topic, and trade and technical. Programme supplements, the most widely read, are published once a week and carried in daily newspapers. Editorial content is largely concentrated on previews of radio and television programmes. Quality supplements are found in some of the major newspapers and are largely read by males in managerial and professional positions. Special topic supplements come out at irregular intervals with various carriers. Trade and technical supplements focus on topics of interest to readers of the specialised journals in which the supplements are placed.

Imagine you are a media buyer and have, say, six different products or services that you want to advertise in German magazines. What products/services would you place where?

3.4.2 Television

Before TV can be included on the media options list, two important questions need to be answered.

- How extensive is television ownership?
- Do commercial stations exist?

In the UK, 98% of households have a television. In India, by comparison, television ownership is low, as purchase of the cheapest set is beyond the means of the majority. However, television watching may be a group activity and so while **ownership** is low, **reach** might be higher.

The opportunities for television advertisers seem overwhelming in Italy, where there are 15 licensed national TV channels and several hundred local TV stations. Advertisers face a different sort of problem in Norway. The first terrestrial commercial channel started broadcasting in September 1992. Before that date, the only television advertising seen by Norwegians was on satellite stations.

Satellite television is perceived to offer good opportunities for reaching wide audiences across national boundaries. MTV, the pan-European music station, targets 16 - 34 year olds and is received in 27 different countries.

3.4.3 Outdoor

The UK lags behind many European countries in its use of posters as an advertising medium. Outdoor offers a good opportunity to communicate to international target groups, as copy tends to be minimal and visuals are key.

3.4.4 Cinema

Cinema may not always be experienced in the same way around the world. In some countries, cinema is viewed in outdoor theatres. There can also be dramatic variations in the quality of the films which are screened. Both these factors may affect how advertising is received by cinema goers.

While the cost of purchasing a television set is prohibitive in some nations, radio ownership seems to be ubiquitous around the world. In most countries, radio plays the role of support medium.

3.4.6 Media planning and buying

As well as the usual media planning considerations (campaign objectives, target markets, media availability, budgets), there may be specific factors which need to be taken into account when planning in international markets. For a food product, media scheduling will need to take account of climatic and seasonal characteristics of individual countries.

Some international print media may be purchased from the home country market via international media representatives. Other media may have to be bought by a local country agent, acting for the advertiser.

Media buying practices can vary from country to country. In the UK, published rate card costs tend to be negotiable downwards. Other countries may be less flexible in negotiations.

Global case study

MTV

MTV potentially faces strong competition from the BBC and Virgin since the launch of digital television. It needs to reinforce its position by raising awareness, but this is a problem because audience perceptions of MTV vary widely across Europe.

MTV's Europe Music Awards ceremony was targeted mainly at the Netherlands, Switzerland and the UK, but research indicated that it needed to adopt different approaches in each country.

(a)   In the UK its tactic was to highlight how many superstars were coming to the awards and play up the competitive aspect. One poster showed stars like George Michael and Bryan Adams; another featured rival Britpop acts Oasis and Pulp.

(b)   In the Netherlands young people like to see MTV as a reassuringly familiar part of their lives. The marketing had a softer more subdued feel, and consists of magazine ads with soft images of domestic situations.

(c)   In Sweden MTV's viewers tend to be more politically disaffected and an anti-establishment image is more appealing. The campaign focused on newspaper ads consisting of replica newspaper pages with graffiti style slogans such as 'Join Our Party'. MTV also placed Join Our Party toilet rolls in the toilets of bars and nightclubs, to achieve the desired subversive effect.

3.5 Legal considerations

Laws and regulations governing marketing communications must obviously be observed. As we saw in Chapter 9, each country will have its own set of restrictions which apply to advertising, packaging, sales promotion or direct marketing.

Direct marketing is an area under threat from EU legislation. In the UK, direct marketers must offer consumers the choice of opting out of receiving further mailings. Recent European proposals suggest the idea of allowing consumers the choice of opting into direct mailings. This seems to imply that companies would have to write to consumers in advance of a direct mail campaign, asking consumers if they wanted to receive mail.

3.5.1 Cross border advertising

The European Commission issued a green paper in May 1996 on *Commercial Communications in the Internal Market*. This covered all forms of advertising, direct marketing, sponsorship, sales promotions and public relations. The paper explained the savings that could be made if advertisers could benefit from economies of scale within the EU by standardising campaigns, as far as cultural differences allow. It anticipated that, without action, the problem would get worse with the advent of the Internet.

The current position is that there must be overriding reasons relating to the **public interest** for imposing additional national rules on communications, and these rules must be 'proportionate' to the public interest objectives. However, to test the validity of current national restrictions on commercial communications through the courts takes several years for each case. Many regard national advertising restrictions as thinly veiled protectionism.

The green paper proposed that a new body be set up as an alternative to the courts. It would be chaired by the European Commission and made up of representatives of each member state, with a brief to assess whether national rules meet the public interest and proportionality tests. There would also be a new central contact, enquiry and information point on commercial communications.

3.6 Standardisation v adaptation

Strategies appropriate for companies operating in international markets

(a)   **Standardise** product/standardise communication (for example, Coca-Cola).

(b)   **Standardise product/adapt communication** (Horlicks is promoted as a relaxing bedtime drink in the UK and as a high protein energy booster in India).

(c)   **Adapt product/standardise communication** (washing powder ingredients may vary from country to country depending on water conditions and washing machine technology. However, the communication message of clean clothes is the same).

(d)   **Adapt product/adapt communication**.

(e)   **Invent a new product** to meet the needs of the market.

Activity 6

As you prepare your *Developing Integrated Communications Strategy* assignment, scan the marketing and advertising trade press alongside the quality papers for examples of products and services which adopt the different strategies listed above. You will be able to use this to help you make decisions for your own organisation.

Advantages of standardising communications include the following.

(a)   **Economies of scale** can be generated. A single worldwide advertising, packaging or direct mail execution will save time and money.

(b)   A **consistent and strong brand image** will be presented to the consumer. Wherever users see the brand, they will be reassured because the messages received will be the same.

(c)   A **standardised communications policy** allows for **easier implementation** and control by management.

(d)   Good **communications ideas are rare** and should be exploited creatively across markets.

Pirelli

A Pirelli ad featured Marie-Jo Perec, a double Olympic gold medal winner, outrunning an avalanche, a tidal wave and a river of volcanic lava. The ad is thought to have cost over £1m to make, but it ran in up to 40 countries from China to South Africa. This needs to be compared with the cost of making multiple commercials, a different one for each country: it is thought to cost at least £250,000 to make a decent 30-second ad.

There are certain clues which showed that this ad was designed to run internationally. There was no dialogue, as that would mean expensive and potentially difficult translation. And the images were universally recognisable – human against nature. The only potential problem was that Perec's sprinting costume could be considered too skimpy for some Islamic countries.

Arguments against standardising communications include the following.

(a) Any standardisation policy **assumes consumer needs and wants are identical** across markets. This may be a false assumption, as the example below illustrates.

(b) Centrally-generated communications concepts may prove to be **inappropriate for the specific culture** of the local market.

(c) **Media channel availability and infrastructure varies** widely from country to country.

(d) A country's level of **educational development** may prevent a standardised approach. For instance, a press campaign featuring detailed copy would be a non starter if literacy levels were low.

(e) **Legal restrictions** may prove to be a stumbling block. For example, France does not allow any advertising of alcohol on television; cashback sales promotion offers are not allowed in Italy or Luxembourg.

(f) Standardisation may encourage the **'not invented here' syndrome**, so that local management becomes lacklustre about creative ideas and communications policies imposed from above.

(g) Different countries have economies which may be much more or much less **developed** than others. Factors that need to be considered are as follows.

- What is the level and trend in per capita income?
- Is the balance of payments favourable or unfavourable?
- Is inflation under control?
- Are the exchange rates stable?
- Is the currency easily convertible?
- Is the country politically stable?
- How protectionist is the country?
- Who controls distribution channels?

3.7 International communications alternatives

It is all too easy to focus on advertising and packaging as the prime communicators in international campaigns. However, sales promotion, public relations, sponsorship, exhibitions, direct marketing and personal selling can all play a part.

3.7.1 Sales promotion

Sales promotion encompasses a range of techniques appropriate for targeting consumers, for instance via price reductions, free gifts, or competitions. Trade and salesforce promotions are also implied under the general heading of sales promotion.

Different countries have their own **local restrictions** concerning different sales promotional devices. For instance, collector devices are a well used method of encouraging repeat purchase in the UK. However, they are not allowed in Luxembourg, Austria, Norway, Sweden and Switzerland. Petrol forecourt promotions are very familiar to drivers in the UK, but following a vicious sales promotions battle between petrol companies in 1989, the Malaysian Government banned sales promotion of petrol.

Sales promotions that are to run across a number of different countries must tap into common tastes, interests and activities. As has been seen from the section on culture, this may not be an easy task.

3.7.2 Public relations

The UK's Institute of Public Relations describe public relations as, 'the planned and sustained effort to establish and maintain goodwill and mutual understanding between an organisation and its publics'. Publics can include:

- Customers
- The local community
- Shareholders
- Media
- Employees
- Government
- Suppliers
- Pressure groups
- Trade intermediaries

The importance of any one group will vary from country to country, which will mean that public relations campaigns are **unlikely to be completely standardised** across markets.

There is some debate as to whether public relations can be standardised at all. There is wide disparity in the sophistication of PR from country to country, as well as the usual list of cultural, language, media and legal barriers. The **political context** of individual countries may also affect the extent to which different public relations techniques can be used.

3.7.3 Sponsorship

Sponsorship seems a particularly appropriate means of communicating internationally. Internationally televised sporting events have an appeal that crosses borders and cultures, as football, rugby and cricket matches often show.

3.7.4 Exhibitions

Trade fairs and exhibitions are an effective means of initial **business-to-business contact**, providing an opportunity for producers, distributors, customers and competitors to meet. Where specialist exhibitions take place, rival companies from across the globe are likely to gather under one roof, offering an excellent opportunity for international buyers to compare offerings and place orders.

Exhibitions allow companies to display goods and mount demonstrations and trials. They can be good as a mechanism for **image building**, especially in new markets. Exhibitions provide a means for supporting local distributors and agents and allow direct contact with customers. There is also the chance to conduct competitor and customer market **research**.

Detailed evaluation of exhibition results should be carried out in order to fine tune future plans.

3.7.5 Direct marketing

International direct marketing has been growing slowly over the last decade. Factors driving the move to international direct marketing include the following.

(a)   The growth in sophistication of computer and database **technology**

(b)   The increasing availability of suitable consumer or business **listings**

(c)   Growth of **international media**, which can be used for direct response advertising

(d)   The perceived **accountability** of direct marketing campaigns compared with other communications campaigns

(e)   The ease with which direct marketing campaigns can be **pre-tested** in order to maximise their effectiveness

(f)   Improving **skills** of direct marketing agencies

(g)   Increasing **willingness of the consumer** to purchase items directly

(h)   Increasing use of **internationally accepted credit cards**

On the other hand, there are a number of factors restraining the move to international direct marketing.

- Lack of telephone and postal **infrastructure**
- Lack of road and rail penetration to facilitate **distribution**
- Lack of **suitable media** to use to target consumers
- Lack of consumer and business **lists** in some countries
- The threat of increasingly **strict legislation** concerning the use of consumer information
- **Consumer backlash** against what is seen as junk mail

3.7.6 Personal selling

Many of the communications tools described above will rely on integration with the personal selling function. For instance, certain types of consumer sales promotion will require **salesforce support** to gain trade acceptance. Likewise, direct marketing campaigns will create sales leads that require follow up. Personal selling is therefore a likely requirement of most markets.

Companies may use **agents or distributors** to act in a sales capacity. Alternatively, a company may establish its own branch offices abroad with a team of local salespeople.

Personal selling will be particularly important in industrial markets where products are high value and complex, or in markets where purchasing is controlled by government agencies.

Assessment advice

Students completing **assignments** integrating international aspects of communications tend to focus on advertising and packaging as the prime promotional vehicles. You need to demonstrate that other options such as sales promotion, public relations, sponsorship, exhibitions, direct marketing and personal selling have been considered – and that these are linked to branding and online activities.

3.8 Multinational communications agencies

The factors influencing the choice of communications agencies were discussed in section 1 of this chapter. In broad terms these factors would also influence selection of agencies to handle **communications on an international basis**. Decisions would also take into account the differences in the kinds of approaches to communications discussed elsewhere in this chapter.

Often the decision making process will be influenced by the structure of the client company and how far they **devolve marketing responsibility** on a regional basis. Some elements of the communications strategy

and branding may be handled centrally, whereas **implementation is dealt with locally** where more detailed knowledge with regard to media and other context issues can be applied.

For many of the larger multinational and global companies such as Coca-Cola, Heinz or Procter and Gamble there is a need to deal with **agencies who have sufficient resources** of their own to handle the size of budgets being allocated. As there has been **consolidation** taking place among multinational businesses through acquisition and merger, this has also been happening among communications businesses. Agencies such as WPP have grown partly by success in winning new business but also through the acquisition of other agencies in different parts of the world.

Other agencies offer multinational services to clients by being part of **networks**. They maintain independence as individual companies but via informal and sometimes formal collaborations, they can provide client services in different geographic markets. Sometimes clients may appoint a **lead agency** who will hold ultimate responsibility for planning and co-ordination of agencies in different markets.

Digital technologies make the process of handling the communications needs of sophisticated multinational clients more straightforward but may not always replace the need for local knowledge.

4 Business to business communications

4.1 Key accounts

So far we have considered the retention of customers as an unquestionably desirable objective. However, for many businesses a degree of discretion will be advisable. 'Key' does not mean large. A customer's potential is very important. The definition of a key account depends on the circumstances. Key account management is about managing the future.

> **Definition**
>
> Key accounts are profitable, and have potential for long-term development

Customers can be assessed for desirability according to such criteria as the profitability of their accounts; the prestige they confer; the amount of non-value adding administrative work they generate; the cost of the selling effort they absorb; the rate of growth of their accounts and, for industrial customers, of the turnover of their own businesses; their willingness to adopt new products; and their credit history. Such analyses will almost certainly conform to a Pareto distribution and show, for instance, that 80% of profit comes from 20% of the customers, while a different 20% generate most of the credit control or administrative problems. Some businesses will be very aggressive about getting rid of their problem customers, but a more positive technique would be to concentrate effort on the most desirable ones.

These are the key accounts, and the company's relationship with them can be built up by appointing key account managers.

Key account management is often seen as a high-level selling task, but should in fact be a business-wide team effort about relationships and customer retention. It can be seen as a form of co-operation with the customer's supply chain management function. The key account manager's role is to integrate the efforts of the various parts of the organisation in order to deliver an enhanced service. This idea has long been used by advertising agencies and was successfully introduced into aerospace manufacturing over 40 years ago. It will be the key account manager's role to maintain communication with the customer, note any developments in his circumstances, deal with any problems arising in the relationship and develop the long-term business relationship.

4.1.1 The key account relationship

The key account relationship may progress through several stages.

(a)   At first, there may be a typical adversarial sales-purchasing relationship with emphasis on price, delivery and so on. Attempts to widen contact with the customer organisation will be seen as a threat by its purchasing staff.

(b)   Later, the sales staff may be able to foster a mutual desire to increase understanding by wider contacts. Trust may increase.

(c)   A mature partnership stage may be reached in which there are contacts at all levels and information is shared. The key account manager becomes responsible for integrating the partnership business processes and contributing to the customer's supply chain management. High 'vendor ratings', stable quality, continuous improvement and fair pricing are taken for granted.

5   Internal communications

Communication within the marketing organisation will be an essential part of an effective IMC approach, ensuring that:

- All units of the organisation who deal with customers (directly or indirectly) are aware of the messages being communicated by themselves and others

- All units understand the desired core messages of the organisation/brand, and the need for consistency in getting that core message across

- Promises made by one customer contact are delivered by subsequent customer contacts!

If internal communications is part of an organisation's strategic communication function, then employees are a vital stakeholder group within the context of communications. The behaviour of employees contribute towards the corporate identity and project it to the external stakeholders, (Yeoman and Tench (2006).

Yeoman and Tench (2006) identified that:

'The strategic purpose of internal communication can perhaps best be summarised as one that is concerned with building two-way, involving relationships with internal publics, with the goal of improving organisational effectiveness'.

Global case study

The Centre for Integrated Marketing Communications specified Harley Davidson as an example of best practice where not only were communications integrated but the entire marketing function was integrated with the rest of the business.

The report authors Jenkinson and Sain (2005) stated:

"Integrated Marketing depends first on uniting everyone around a collective vision of value that connects to the identity and purpose of the organisation/brand. This depends on a profound and shared understanding of customers and an organisation that can deliver value seamlessly throughout all customer experiences across the relationship. This also means connecting and matching spiritual with practical qualities: vision, purpose, values with information, processes, and systems." p.3

The motorcycle manufacturer had experienced a period of strategic nightmare. In the 10 years to 1983, Harley's market share of the 850 CC plus motorcycle category had dropped from 80% to 23%. The company was haemorrhaging cash and profits. Staff were demoralised. The culture and environment were toxic.

Integrated Marketing amounts to a widening of the responsibility, potential and vision for many marketers and therefore for marketing. At Harley Davidson a number of initiatives were used over a number of years to regain strength within the company. The authors reported that:

"*Involvement, empowerment and alignment was the secret of success. The result is something called the Business Process, an extensive and ongoing programme that involved and involves everyone in the organisation from top to bottom in establishing shared values and vision, shared mission and operating philosophies, and agreed objectives and strategies. The extensive change process, over several years, that led to this, known as the Joint Vision Process, also led to a radical new organisation. Instead of the conventional hierarchical structure, Harley Davidson developed what they call a circle organisation of three overlapping elements concerned with creating demand, producing products and providing support. A leadership and strategy council at the centre has members nominated from these circles.*"

The case study is available to download from:

http://www.centreforintegratedmarketing.com/gfx/documents/harley_davidson_-_organisation-led_im.pdf

5.1 The changing role of internal communication

It is important to note that the role of internal communications has changed significantly and it is now considered to be an important part of strategic communications at a corporate level. Historically, internal communications was not viewed in this way and this is a significant shift of mindset. It has been recognised that internal communications contributes towards the bottom line and directly impacts upon the reputation of the organisation. As a result, then of the significance of internal communications, it is critical not only to provide information but to provide a context for that information as well. Doorley and Garcia (2007) citing a director of employee communications Barry Mike, describes eight business trends that have helped set the scene for an improved employee communication function:

1   **Growing cynicism** – as a result of greedy CEOs and an enlightened attitude towards employee communications.

2   **Organisations have dramatically flattened** – decreasing bureaucracy and bringing decision-making closer to the 'front line' employees.

3   **The rate of change is increasing** – 'by the book' mentality is no longer valid. Emphasis is now on flexibility and elasticity.

4   **Speed has become a differentiator** – because of information technology – rather than just price and quality differentiating products.

5   **Lifetime employment is dead** – there is no longer the prospect of life long employment in return for loyalty, hence all employees can be offered is the prospect of expanding into new roles.

6   **Human capital** – that is the people becoming the key competitive differentiator.

7   **The desire for purpose** – employee communications can help employees feel part of 'something bigger'.

8   **Service sector rise** – resulted in internal branding being more important – as employee communications has become a vital extension of marketing and strategic planning in ensuring the delivery of brand promise.

Krone et al, (2001) suggests that there are four roles which internal communications should fulfil:

- Efficiency
- Shared meaning
- Connectivity
- Satisfaction

5.1.1 Efficiency

Internal communications is fundamentally used to communicate and disseminate information about the corporate activities of the organisation. As an important stakeholder group, employees need to be fully informed, whatever the communication.

Efficiency means functioning effectively with little wasted effort (Collins, 2006, p. 88). Efficiency in terms of internal communications means that communications among employees is effective and efforts are not wasted, resources being used carefully and thoughtfully, without waste. The efficiency of internal communications can be measured to evaluate how efficient it has been.

Lyn Smith in her book Internal Communications reports on the UK's CIPR 2002 Excellence Award winner, Seeboard. Seeboard was highlighted for conducting an evaluation at every stage of its branding project, using questionnaires, surveys, intranet surveys and mystery shopper-type surveys where performance was tested without the employees being aware of the process. An example of the evaluation was the first phase survey which showed that 99% of the staff understood the new customer vision. In addition, the twice annually staff survey recorded an 83% response rate. Seeboard was reported in this survey as being:

- A good place to work 88%
- Felt proud to work for Seeboard 92%
- An amazing 92% said they knew the priorities of their team.

Smith suggests that it is useful to look backward as well as onward in terms of measuring and planning for efficiency.

5.1.2 Shared meaning

The focus of internal communications has changed from addressing events and people to sharing corporate goals (Smith, 2009). Internal communications can now be used as a vehicle for building a shared understanding with employees about corporate aims and goals. It is also important to consider different perspectives and what it is like to be on the receiving end. What are the staff perceptions, for example?

5.1.3 Connectivity

In order to achieve the high level of performance so many companies seek, it is necessary to show the connection between the company's success and the success of its employees. All need to understand what needs to be achieved whatever role they play in that process in order that they understand the 'bigger picture' (Quirke, 2008).

Once a message has been planned and sent adding the opportunity to engage and express opinion opens up a virtuous circle of communication. Ideas and suggestions are fed back into the system with the intention of keeping communications flowing and 'connecting'. Smith (2008) highlights that **more than 50% of communication is listening**. One needs to ensure that listening is taking place or otherwise even the best designed and developed communication will be worthless.

5.1.4 Satisfaction

Internal communications is also used to improve job fulfilment throughout the organisation. If employees are satisfied, they perform better and are more motivated. Happy employees create a pleasant working environment.

5.2 Planning internal communication

A very wide variety of communication media is used in formal internal communications, including: company reports, brochures and newsletters; corporate intranet (staff website); video- and web-conferencing; video and CD-ROM presentations and seminars; team meetings, briefings, presentations and conferences; one-to-one interviews; negotiating and consultative meetings (eg with employee representatives); letters, e-mail, memoranda and notices. Both formal and informal channels are important. It is essential that in order to maintain positive relationships the attempts made at communication are effective and measured.

5.2.1 Formal communication channels

Formal communication channels are those which exist within the organisational structure. How complex these communication structures are depends very much on the complexity of the origination concerned in terms of how bureaucratic it is (in terms of process, policy and procedure for many things but certainly for communications); how hierarchical it is (how many layers there are of people within the organisational structure) or how flat the organisation is.

How an organisation is structured has a significant impact upon how communication flows.

Quirke (2008) argues that communication channels need to be better managed:

"...to target information, reduce interruptions by irrelevant messages, and liberate employee's time and 'brain space".

Organisations need the right mix of communication channels and to be able to use the right channel for the right type of information. Equally, organisations need to target their audiences more closely and tailor messages more relevantly to address their peoples' interests and concerns. (Quirke, 2008).

5.2.2 Informal communication

Informal communication occurs within the informal organisation – an informal organisation always exists alongside the formal one. This was covered in detail in Chapter 3. Unlike the formal organisation, the informal organisation is loosely structured, flexible and spontaneous. It embraces such mechanisms as:

- **Social relationships** and groupings (eg cliques) within – or across – formal structures

- The **'grapevine'** or informal communication which by-passes the formal reporting channels and routes

- **Behavioural norms** and ways of doing things, both social and work-related, which may circumvent formal procedures and systems (for good or ill)

- **Power/influence structures**, irrespective of organisational authority: informal leaders are those who are trusted and looked to for advice.

5.3 Effective communication

Effective communication means: the right person receives the right information in the right way at the right time. The focus here is on effective internal communications.

What does 'good communication' look like? It is perhaps easiest to identify poor or ineffective communication, where information is not given; is given too late to be used; is too much to take in; is inaccurate or incomplete; is hard to understand. To clarify, then, effective communication is:

(a)   **Directed to appropriate people**: this may be defined by the reporting structure of the organisation, but it may also be a matter of discretion, trust and so on.

(b)   **Relevant to their needs**: not excessive in volume (causing overload); focused on relevant topics; communicated in a format, style and language that they can understand.

(c)   **Accurate and complete** (within the recipient's needs): information should be 'accurate' in the sense of 'factually correct', but need not be minutely detailed. In business contexts, summaries and approximations are often used.

(d)   **Timely**: information must be made available within the time period when it will be relevant (as input to a decision, say).

(e)   **Flexible**: suited in style and structure to the needs of the parties and situation. Assertive, persuasive, supportive and informative communication styles have different applications.

(f)   **Effective in conveying meaning**: style, format, language, and media all contribute to the other person's understanding or lack of understanding. If the other person doesn't understand the message, or misinterprets it, communication has not been effective.

(g)   **Cost-effective**: in business organisations, all the above must be achieved, as far as possible, at reasonable cost.

5.4 Tools of internal communication

A couple of important points before we discuss the plethora of tools available to internal communications. First, communicators must be skilled at selecting exactly the right media to gain the greatest possible amount of message penetration across the target audience. What are your audiences' communication preferences – how do they want to be communicated with? Second, that despite all advances in technology, that face-to-face communication is still the preferred medium by most employees, so surveys tells us (Smith, 2009).

Here follows some guidance points on internal communications (adapted from Smith 2009, pp. 79–92)

- Focus on the message before thinking about the channels.
- Think – how can I segment the audiences and what are their preferences – in terms of media.
- Consider a 'pick and mix' approach to media not choosing one but a selection.
- Employees rate face-to-face highest on their list – keep message short and simple.
- Use support materials in group face to face meetings.
- Remember to build in two-way feedback mechanisms into the communications.
- Remember that print is easily portable and is a medium of record.
- Keep a balance of news and features in print.
- Use the moving image to engage people.
- Remember that DVDs offer the opportunity for participation.
- Audio cassettes are still good for mobile audiences.
- Email and intranet can be 'hard on the eyes'.
- Adapt material for screen-based media.
- Flashes and alerts can be communicated well by text.
- Some quirky gimmicks like board games can engage people.
- Corporate Social Responsibility (CSR) can motivate the workforce.

It is critical to ensure that investment is being made in communication channels which demonstrate a clear return on investment. So often communications teams operate a range of channels which in practice actually duplicate effort and coverage and don't always use relevant content. It is also critical to understand what communications channels are used for what types of information and where, so it is clear where one's effort should be focused.

The following methods are frequently used.

5.4.1 Email

Email has replaced letters, memos, faxes, documents and even telephone calls – combining many of the advantages of each medium with new advantages of speed, cost and convenience. Email offers many advantages for internal customer communication.

- Messages can be sent and received very fast (allowing real time message dialogue).

- Email is economical, especially for international communication: often allowing worldwide transmission for the cost of a local telephone call.

- All parties can print out a 'hard copy' of the message, for detailed perusal and repeated reference.

- Messages can be sent worldwide at any time: email is 24-7, regardless of time zones and office hours.

- The user can attach complex documents (spreadsheets, graphics, photos) where added data or impact are required.

- Email message management software (such as Microsoft Outlook) has convenient features such as: message copying (to multiple recipients); integration with an 'address book' (database of contacts); corporate stationery options; and facilities for mail organisation and filing.

Activity 7

From your knowledge of email, what might be some of the disadvantages of email for internal communication with colleagues in your own department or other departments.

5.4.2 Intranet

An intranet is an internal, mini version of the Internet, using a combination of networked computers and web technology. 'Inter' means 'between' and 'intra' means 'within'. This may be a useful reminder. The Internet is used to disseminate and exchange information among the public at large, and between organisations. An intranet is used to disseminate and exchange information within an organisation: only employees are able to access this information.

The corporate intranet may be used for:

- **Performance data**: linked to sales, inventory, job progress and other database and reporting systems, enabling employees to process and analyse data to fulfil their work objectives

- **Employment information**: online policy and procedures manuals (health and safety, equal opportunity, disciplinary rules, customer service values and so on), training and induction material, internal contacts for help and information

- **Employee support/information**: advice on first aid, healthy working at computer terminals, training courses offered

- **Notice boards** for the posting of messages to and from employees: notice of meetings, events, trade union activities

- **Departmental home pages**: information and news about each department's personnel and activities, to aid crossfunctional understanding

- **Bulletins or newsletters**: details of product launches and marketing campaigns, staff moves, changes in company policy – or whatever might be communicated through the print equivalent, plus links to relevant pages for details

- **Email facilities** for the exchange of messages, memos and reports between employees in different locations

- **Upward communication:** suggestion schemes, feedback questionnaires, employee attitude surveys

- **Individual personnel files**, to which employees can download their training materials, references, certificates, appraisals, goal plans

- **Social notices**: including sale of items, social groups.

Advantages cited for the use of intranets include the following.

- **Cost savings** from the elimination of storage, printing and distribution of documents that can instead be exchanged electronically or be made available online

- **More frequent use made of online documents** than printed reference resources (eg procedures manuals) and more flexible and efficient searching and updating of data

- **Wider access to corporate information.** This facilitates multi-directional communication and co-ordination (particularly for multi-site working). It is also a mechanism of internal marketing and corporate culture.

- **Virtual team working.** The term 'virtual team' has been coined to describe how ICT can link people in structures which simulate the dynamics of team working (sense of belonging, joint goals, information sharing and collaboration despite the geographical dispersion of team members in different locations or constantly on the move

5.4.3 Team meetings

Meetings play an important part in the life of any organisation.

- **Formal discussions** are used for information exchange, problem solving and decision making: for example, negotiations with suppliers, meetings to give or receive product/idea presentations or 'pitches', and employee interviews.

- **Informal discussions** may be called regularly, or on an ad hoc basis, for communication and consultation on matters of interest or concern: for example, informal briefings and marketing or project team meetings.

Activity 8

Team meetings may appear to be relatively inconvenient, especially when compared to modern communications such as group emails. Why do you think that meetings are still held, and are still useful?

5.4.4 SMS

SMS text messaging commonly used in advertising and direct marketing media. However, it is also a common tool of quick messaging in a variety of communication contexts, including internal. It is quicker and less intrusive than a phone call (since it does not need to be accepted or returned immediately), and enables mobile messaging for those without immediate access to email (although SMS messages can also be sent from phones to computers and vice versa).

Karlshamns Oljefabriker

Karlshamns Oljefabriker is an oil-based food product company. At Christmas all employees were given a variant of Trivial Pursuit, one of the world's most loved quiz games. The mission of the factory of only supplying consumer co-operative stores had been broadened to include the sales of its products to all types of retailers. A major organisational audit had been performed and the problems of informing everyone were massive. Of the 2,000 questions in Trivial Pursuit, 500 were substituted. The new questions were of local nature, both of general interest such as 'Which year were the first winter Olympics held?' to direct questions regarding the company, for example, ... 'Which are the two most important ingredients in our margarine?'

The game was a hit. Employees played it over Christmas, in their spare time. That was also the idea; it was handed out during the last hours before the Christmas holiday. This way, its messages also reached family members and friends. Others who did not work in the factory asked for the game after it made the news in the local paper and radio station. So the internal marketing activity reached beyond the staff and had a positive mega communications effect, particularly in the home town of the factory.

(Source: Gummesson, 1999, p. 163)

5.5 Barriers to internal communication

Barriers to communication include 'noise' (from the environment), poorly constructed or coded/decoded messages (distortion) and failures in understanding caused by the relative positions of senders and receivers. General problems which can occur in the communication process include:

- Distortion and noise factors (described above)
- Misunderstanding due to lack of clarity or technical jargon
- Non-verbal signs (gesture, facial expression) contradicting the verbal message
- Failure to give or to seek feedback
- 'Overload' – a person being given too much information to digest in the time available
- Perceptual selection: people hearing only what they want to hear in a message
- Differences in social, racial or educational background
- Poor communication skills on the part of sender or recipient

Additional difficulties may arise from the **work context**, including:

(a) **Status** (of the sender and receiver of information)

    (i) A senior manager's words are listened to closely and a colleague's perhaps discounted.

    (ii) A subordinate might mistrust his or her superior's intentions and might look for 'hidden meanings' in a message.

(b) **Jargon**. People from different job or specialist backgrounds (eg accountants, HR managers, IT experts) can have difficulty in talking on a non-specialist's wavelength.

(c) **Priorities**. People or departments may have different priorities or perspectives so that one person places more or less emphasis on a situation than another.

(d) **Selective reporting**. Subordinates may give superiors incorrect or incomplete information (eg to protect a colleague, to avoid 'bothering' the superior). A senior manager may, however only be able to handle edited information because he does not have time to sift through details.

(e) **Use**. Managers may be prepared to make decisions on a 'hunch' without proper regard to the communications they may or may not have received.

(f)   **Timing**. Information which has no immediate use tends to be forgotten.

(g)   **Opportunity**. Mechanisms, formal or informal, for people to say what they think may be lacking, especially for upward communication.

(h)   **Conflict**. Where there is conflict between individuals or departments, communications will be withdrawn and information withheld.

(i)   **Cultural values** about communication.

    (i)   **Secrecy**. Information might be given on a need-to-know basis, rather than be considered as a potential resource for everyone to use.

    (ii)   **Can't handle bad news**. The culture of some organisations may prevent the communication of certain messages. Organisations with a 'can-do' philosophy may not want to hear that certain tasks are impossible.

    (iii) **Religion**. Religion can block messages or change how messages are interpreted.

Smith (2009) in her book *Internal Communications*, summarises some typical blocks to internal communication as follows.

- Age
- Gender
- Disability
- Culture (religion)
- Channel distortion
- Overactive grapevine
- Previous history of organisation
- Distrust in management
- Too much of the same thing
- Regional differences

5.6 Improving downward communication

Depending on the problem, measures to improve communication may be as follows.

(a)   **Encourage, facilitate and reward** communication. Status and functional barriers (particularly to upward and inter-functional communication) can be minimised by improving opportunities for formal and informal networking and feedback.

(b)   **Give training and guidance** in communication skills, including consideration of recipients, listening, giving feedback and so on.

(c)   **Minimise the potential for misunderstanding** by making people aware of the difficulties arising from differences in culture and perception, and teach them to consider others' viewpoints.

(d)   **Adapt technology, systems and procedures** to facilitate communication: making it more effective (clear mobile phone reception), faster (laptops for emailing instructions), more consistent (regular reporting routines) and more efficient (reporting by exception).

(e)   **Manage conflict and politics** in the organisation, so that no basic unwillingness exists between units.

(f)   **Establish communication channels and mechanisms** in all directions: regular staff or briefing meetings, house journal or intranet, quality circles and so on. Upward communication should particularly be encouraged, using mechanisms such as inter-unit meetings, suggestion schemes, 'open door' access to managers and regular performance management feedback sessions.

Communication between superiors and subordinates will be improved when **interpersonal trust** exists. Exactly how this is achieved will depend on the management style of the manager, the attitudes and personality of the individuals involved, and other environmental variables. Peters and Waterman advocate 'management by walking around' (MBWA), and **informality in superior/subordinate relationships** as a means of establishing closer links.

5.7 Improving upward communication

Consultation is particularly important, as a mechanism for upward communication.

Consultation is, basically, a process of gathering information and views from another party before making a decision.

(a) **Formal consultation** is required in some business contexts. Employers must consult with elected employee representatives on issues of concern to them and their work, including re-structuring and proposed redundancies.

(b) **Informal consultation** is a managerial style in which the manager invites team members to contribute their views and input to decisions.

5.8 Improving lateral communication

In addition to addressing upward and downward communication between employees and managers, it is also important to improve communications laterally – that is between employees working at similar levels.

The best practices in employee communications are summarised here taken from Doorley & Garcia (2007).

- The best stick to classic PR process: research, planning, communication and evaluation.
- Lead employee communicator has direct access to the senior leadership team.
- Employee communicators are senior by leadership as trusted advisers.
- Communicators articulate a communications strategy that is directly linked to business objectives.
- The communicators inventory, profile and prioritise their employee audiences.

http://www.crainc.com/commlog. A web log dedicated to the communications strategy and practice inside organisations.

5.9 The context of internal communications

There are some other aspects of internal communications that one needs to consider. For example, it is important to have some familiarity with what people actually want to hear and what subjects interest them. The following table, adapted from Smith (2008) summarises subjects that are of interest to employees and rates them on a scale of 1–10.

SUBJECTS OF INTEREST TO EMPLOYEES		
Rank	Subject	Scale 1–10
1	Organisational plans for the future	8
2	Job advancement opportunities	7
3	Job-related how-to information	7
4	Productivity improvements	6
5	Personal policies and practices	6
6	How we're doing v the competition	6
7	How my job fits into the organisation	6
8	How external events affect job	5
9	How profits are used	5
10	Financial results	4

One other useful point here is the four-step process of a conversation, according to Quirke (2008, p. 177) which includes:

- **Content** – keep messages simple so that the meaning is clear.

- **Context** – provide a context within which the communication takes place – so that they can relate to what is being said.

- **Conversation** – interaction, skills and dialogue are essential to the conversation process.

- **Feedback** – what feedback is required – something that needs due consideration. Receipt of messages, such as satisfaction or ideas for improvements?

It is also worth considering the breadth of internal communications that takes place. Smith (2008) cites a spectrum of internal communication, originally developed by Russell Grossman, then Head of Internal Communications for the BBC.

(a) **Internal communication spectrum**

   (i) **Day to day communication**

- Organisation news
- Awards, etc
- Housekeeping – pay, appraisals, appointments, operational, etc
- Interpreting external media coverage
- Listening and responding to worries, etc

   (ii) **Change communication**

- Explaining vision, purpose and values
- Helping leaders talk with their teams
- Sharing knowledge and best practice
- Embedding change initiatives
- Listening for reactions
- Understanding responses to change
- Celebrating team success
- Industry and other 'bigger picture' news

   (iii) **Marketing communication**

- Creating passion about product
- Internal marketing activity
- Inviting critique and discussion
- Marketing programmes
- Visual environment (building etc)

Assessment advice

Ensure that you have conducted a thorough review of the internal marketing within your own organisation.

1 Managing agencies and implementing communication plans

- Implementation requires careful management using project management skills

- A number of barriers to implementation exists

- Stakeholders need managing to encourage shared vision

- There are a number of reasons why a company might want to use **external marketing services**, rather than plan or carry out marketing activities in-house.

- There are a wide variety of different agency types.

- The brief is a key document outlining client requirements and begins the agency recruitment process

2 Providing consistency with various global audiences

- The initial decision is whether to globalise or not.

- Various cultural dimensions exist

- Countries may differ according to their levels of marketing communications sophistication

- Marketing communications tools may need varying according to different markets familiarity.

3 The difference with business to business communications

- **Key account management** is often seen as a high-level selling task, but should in fact be a business-wide team effort about relationships and customer retention. It can be seen as a form of co-operation with the customer's supply chain management function

4 How to manage internal communications

- Communication within the marketing organisation will be an essential part of an effective IMC approach

- Communicators must be skilled at selecting exactly the right media to gain the greatest possible amount of message penetration across the internal target audience

- If internal communications is part of an organisation's strategic communication function, then employees are a vital stakeholder group within the context of communications.

1 Market Research Limited will provide retailers of different kinds of products with useful information as well as the product manufacturers, most likely consumer goods given this description. Children and Youth – companies producing goods and services aimed at the youth market, who want to conduct research but may not have the resources to undertake surveys independently.

2 **Budget**

Client and agency need to be clear about what the budget covers. For example, how many new photographs need to be shot? This will have a bearing on the production costs. The budget for media will determine the choice for media selection. The agency can alter the resources according to the job.

Timetable

Important key deadlines must be listed. The agency should draw up a timetable for presentation of design concepts,proofs, final proofs and so on.

Objectives

The primary objective may be to increase sales. This can be incorporated in the design.

Target audience

The agency needs a full description of the characteristics of the audience. This will enable them to design the campaign with the audience in mind.

Product history, features and benefits

For example, is this a complete redesign, or just a freshen up?

Be clear what the USP is. For example, you cannot have high quality, cheap products, top brands and commodity goods all in the same catalogue. With a clear picture, the agency will be able to focus on producing what is required without constantly referring back to the client.

3 Ideally, the agency should be able to assure the client of experience in both consumer market generally and the toiletries market in particular. The agency should have local offices in a number of the Middle East states in which the client does business and be used to handling local and international campaigns in those countries. The agency must demonstrate knowledge concerning cultural, legal, and media difficulties which may be encountered and be able to propose solutions for overcoming these problems. Proof of successful campaign outcomes for other British based clients advertising in related markets would also be reassuring.

4 This will depend on your own research.

5 This will depend on your own research.

6 This should be an ongoing activity that you integrate into your module work.

7 This will depend on your own organisation.

8 It is essential to develop effective working relationships. In today's virtual world, it is valuable to "touch base" from time to time.

Collins English Dictionary (2006). HarperCollins Publishers.

Doorley, J. & Garcia, H.F., (2007). *Reputation Management: The Key to Successful Public Relations and Corporate Communication.* New York: Routledge.

Gummesson, E., (2002). *Total Relationship Marketing.* 2nd edition. Oxford: Elsevier Butterworth-Heinemann, Oxford.

Jenkinson, A., Sain, B. (2005). *Integrated Marketing: a New Vision, in Marketing Mind Prints* edited by Philip Kitchen. Basingstoke UK/New York N.Y: Palgrave-Macmillan.

Jobber, D., (2007). *Principles and Practice of Marketing.* 5th edition, Maidenhead: McGraw-Hill Education.

Pickton, D and Broderick, A., (2004). *Integrated Marketing Communications.* 2nd edition. Harlow: Prentice Hall.

Smith, L., (2008). *Effective Internal Communication.* 2nd edition. London: CIPR Practice Series, Kogan Page Limited.

Smith, P.R., & Taylor, J., (2004). *Marketing Communications: an Integrated Approach.* 4th edition. London: Kogan Page.

Quirke, B., (2008). *Making the Connections: Using Internal Communication to Turn Strategy into Action* 2nd edition. Aldershot, Hampshire: Gower Publishing Limited.

Van Riel, C.B.M. & Fombrun, C.J., (2007). *Essentials of Corporate Communication: Implementing Practices for Effective Reputation Management.* Abingdon, Oxon: Routledge.

Tench, R. & Yeoman, L., (2006). *Exploring Public Relations.* Harlow, Essex: Pearson Education Limited.

 Developing Integrated Communications Strategy

Notes